Naturalists' Handbooks 19

Pollution monitoring with lichens

D.H.S. RICHARDSON

With colour plates by Claire Dalby

T0308376

Pelagic Publishing
www.pelagicpublishing.com

Published by **Pelagic Publishing**
www.pelagicpublishing.com
PO Box 874, Exeter, EX3 9BR, UK

Pollution monitoring with lichens
Naturalists' Handbooks 19

Series editors
S. A. Corbet and R. H. L. Disney

ISBN 978-1-78427-211-1

Digital reprint edition of ISBN 0-85546-289-2 (1992)

Text © Pelagic Publishing 2019
Illustrations © Claire Dalby 1981

All rights reserved. Apart from short excerpts for use in
research or for reviews, no part of this document may
be printed or reproduced, stored in a retrieval system,
or transmitted in any form or by any means, electronic,
mechanical, photocopying, recording, now known or hereafter
invented or otherwise without prior permission from the
publisher.

British Library Cataloguing in Publication Data
A catalogue record for this book is available from the British
Library.

Contents

Editors' preface

Author's preface and acknowledgements

1	Introduction	1
2	Natural history	3
3	Sulphur dioxide and acid rain	6
4	Ozone and nitrogen compounds	16
5	Fluorides	18
6	Aromatic hydrocarbons	22
7	Metals	24
8	Transplant studies	34
9	Radioactive elements	37
10	Invertebrate fauna	40
11	Identification	43
12	Original work: techniques and approaches	57
	References and further reading	66
	Useful addresses	74
	Index	75

Editors' preface

Students at home or university, and others without a university training in biology, may have the opportunity and inclination to study local natural history but lack the knowledge to do so in a confident and productive way. The books in this series offer them the information and ideas needed to plan an investigation, and practical guidance to help them carry it out. They draw attention to regions on the frontiers of current knowledge where amateur studies have much to offer.

Those amateurs and professionals who have already discovered the fascination of lichens can easily find material for study throughout the year. The growing concern over threats to our environment has brought lichens to the attention of a wider public, and the success of the survey conducted by Irish schoolchildren, described in chapter 3, shows how valuable lichens can be in the detection and mapping of pollution. More studies of that kind are needed. Lichens already have a following of devotees. We hope this book will introduce lichens to a wider audience and enable more people to interpret the information that lichens can provide.

We are very grateful to Claire Dalby and to the publishers for allowing us to re-use her superb coloured illustrations, originally published by the British Museum (Natural History) and BP Educational Service as a chart entitled *Lichens and air pollution*.

S.A.C.
R.H.L.D.
October 1991

Author's preface and acknowledgements

Professor D.L. Hawksworth and Dr F. Rose published their short influential book on *Lichens as Pollution Monitors* in 1976 but it has been out of print for some years. In the meantime, several books (listed on p. 66) have appeared on biological monitors, but none deals exclusively with lichens. In recent years more than a thousand research studies have been completed demonstrating the value of lichens for monitoring purposes. These are listed in the compendium 'Literature on Air Pollution and Lichens' which appears regularly in the international journal, *The Lichenologist*. It is thus timely to present an up-to-date account of the subject although in a short book it is impossible to refer to all the research papers mentioned above. My aim has been to select studies from different parts of the world that illustrate the variety of possible approaches using lichens as pollution monitors. An attempt has been made to describe these studies in the context of the underlying science. I thank Trinity College Dublin, Ireland for approving a short sabbatical during which this book was written and Dr I. Brodo and the Canadian Museum of Nature, Ottawa, Canada, for providing a room, computer facilities and a congenial and hospitable atmosphere in which to write.

I gratefully acknowledge the help of Professor M.R.D. Seaward, Ms S. Pappin and Dr P.J. Beckett who supplied information and unpublished material. I would also like to thank Dr I. Brodo, Professor M.R.D. Seaward, Dr S. Corbet and my wife Roxanne for their extremely valuable comments on part or all of the the first draft of the manuscript and Mr F. Dobson for help in revising the lichen identification key and adding the valuable marginal drawings, one of which, no. 4, was prepared by Ms E. Jennings. Finally, I am grateful to the authors and publishers of various research papers, for permission to reproduce their illustrations in this book, to Alison Skingsley for drawing figs 20 and 26, and particularly to Claire Dalby for the colour plates.

D.H.S.R.
School of Botany, Trinity College, University of Dublin
1991

1 Introduction

In unpolluted areas, the growth of lichens on tree trunks and rocks is usually conspicuous, forming grey, green or even orange patches. Soon after the advent of the Industrial Revolution in Europe, naturalists began to observe that lichens were not found where air was impure (Turner & Borrer, 1839)*. As early as 1859, Grindon attributed the decline of lichens around cities to air pollution. He wrote 'The quality [of lichens near Manchester] has been much lessened of late years ... through the influx of factory smoke which appears to be singularly prejudicial to these lovers of pure air.' However, it was Nylander in the latter half of the last century who realised, after studying the lichens around Paris, that these plants might be useful indicators of air quality. He wrote that lichens 'donnent à leur manière la mesure de solubrité de l'air et constituent une sorte d'hygrometre très sensible'. His studies in the Jardin du Luxembourg showed first the decline and then the disappearance of lichens on trees in the garden (Nylander, 1866, 1896). It is interesting that now, after almost a century, lichens are returning to colonise these trees following a fall in air pollution (Seaward & Letrouit-Galinou, 1991).

In spite of Nylander's pioneer research, it was not until the work of Skye (1958) that tolerance limits of particular species began to be documented. By 1970, Hawksworth and Rose were able to establish a qualitative scale for estimating sulphur dioxide air pollution using lichens growing on tree bark. They examined the lichen flora around cities in England and Wales and related the distribution of particular lichen species to measured winter mean sulphur dioxide levels. They defined a series of ten zones with zone 1 having no lichens ('a lichen desert') and mean winter sulphur dioxide (SO_2) levels in excess of 170 micrograms per cubic metre of air ($\mu g\,m^{-3}$), and zone 10 with a wide range of lichens typical of ancient woodlands and less than 10 $\mu g\,m^{-3}$ of SO_2. Since the publication of this scale, surveys of lichen communities have been used to assess air quality in and around many cities throughout the world.

The tolerant lichen species that survive in polluted areas also provide another type of information. Lichens accumulate substances from rain and trap airborne particles impinging on them. As a result, some pollutants become concentrated in the lichens and chemical analysis of lichen samples can be used to determine the nature of the emissions or to estimate the size of the contaminated area around a particular

* References cited under the authors' names in the text are given in full in Further Reading on p. 66.

industrial installation. This ability to accumulate substances from the air has also been exploited for geobotanical prospecting and for monitoring the fallout of radionuclides from atmospheric nuclear testing or from the accident at the nuclear powered electricity generating plant at Chernobyl in the Ukraine. Furthermore, with the help of transplant studies in which healthy lichens are moved to polluted areas, it is possible to monitor the appearance of damage symptoms or changes in chemical composition and use these to help establish patterns of air quality. The aim of this book is to summarise the ways in which lichens may be used to monitor pollution, and to describe the techniques that have been applied in such studies. It is, however, first necessary to understand something of the natural history of lichens and to know a little about how they are identified.

2 Natural history

Lichens can be found from extreme low tide level on the sea shore to the tops of high mountains, and from arctic to tropical regions. This wide distribution is the more remarkable because a lichen is a symbiotic association between two quite different organisms: a photosynthetic green alga, or less often a cyanobacterium ('blue-green alga'), and a fungus. The photosynthetic units provide carbohydrates and sometimes other compounds to the fungus. The whole association grows at a rate ranging from one millimetre or less per year for crust (crustose) lichens up to a few centimetres a year for the most rapidly growing leafy (foliose) or shrubby (fruticose or pendant) lichens. The larger lichens in the last two categories grow, on average, about 5 mm per year in length or radially and are usually attached to their substratum by complex fungal strands called rhizinae. However, unlike flowering plants, they possess no true roots and are not covered by a protective cuticle. Indeed, lichens have evolved a strategy for survival that is quite different. They quickly absorb water from rain or dew and become metabolically active within a few minutes, although it may take longer for photosynthesis to reach optimal levels (Kershaw, 1985). Under sunny conditions they lose water and can become dry and crisp in less than an hour. As their water content falls, photosynthesis ceases and respiration then stops, after which they remain inactive until remoistened.

Lichens have also evolved very efficient mechanisms for accumulating nutrients from the environment in which they live. For example, they have active uptake systems for anions like nitrates and sulphates. This means that the plant expends metabolic energy to take in these anions which are accumulated within the cell. Lichens adsorb metal ions such as Ca^{2+} via an ion exchange mechanism and can trap tiny particles of rock, soil or pollutants within their structure. Some of the metabolites produced by lichens can break down such particles, releasing nutrients which may then be taken up into the cells of the lichen.

Lichens can colonise almost any surface that does not flake off or erode too quickly, so long as there is enough light for the contained alga or cyanobacterium to photosynthesise. There must also be sufficient moisture for the lichen to stay metabolically active for significant periods during the day or year to enable the lichen to grow in size. Some lichens, about 10% of all species, contain cyanobacteria either as the main photosynthetic partner or in small packets within the lichen where a green alga is also present. The cyanobacterium is able to fix nitrogen from the air and much of this passes to the fungal partner, providing it with a source of nitrogen that supplements the small amount available in rainwater.

The fungal partner of most lichens belongs to a group known as the Ascomycetes and reproduces by ejecting spores from sacs (asci) within fruit bodies that form on the surface or edge of the lichen. Such spores, if they land on a suitable substratum, can germinate and entrap nearby algae or cyanobacteria. If these are compatible photosynthetic partners, a lichen resynthesis will occur and a lichen plant, referred to as a thallus, will begin to grow (Laundon, 1986). Fortunately, many lichen fungi do not need to go through this resynthesis process every time they reproduce. They have evolved vegetative means of propagation in which both partners are distributed together from the parent plant. Some species produce small granules, soredia, which are distributed by wind and rain. Others have developed fragile cigar- or bun-shaped outgrowths, isidia, that break off and can grow into a new lichen on a suitable surface (fig. 1).

Lichens grow relatively slowly and persist for tens or hundreds of years on their substratum, whether it be tree bark or rock. They are nutrient sources for a wide variety of invertebrates which are often specialised lichen consumers. These include protozoa, mites, snails and caterpillars. In some parts of the world, the larger ground- and tree-dwelling lichens form a food source for various deer, especially reindeer (Richardson, 1975, 1991). Certain fungi have evolved to parasitise or grow within lichens as parasymbionts. A variety of aromatic secondary metabolites called lichen substances are synthesised by lichens in amounts from less than 1% up to about 10% of the dry weight. These are often antibiotic in nature and probably help to deter invading microorganisms. Many of the substances also seem to be distasteful to browsing animals, large or small (Lawrey, 1984).

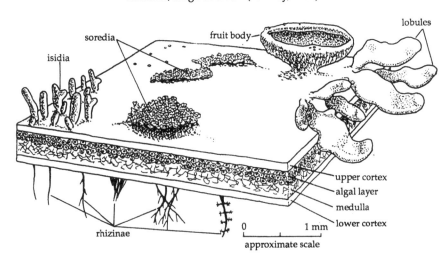

Fig. 1. The structure of a generalised foliose lichen (after Brodo, 1988).

The most basic approach to the identification of lichens involves the use of guides with coloured illustrations (such as Jahns, 1983) or coloured chartlets of which there are two, one dealing with lichens on trees and the other with lichens on the sea shore (Dalby, 1981, 1987). For precise identification, it is necessary to examine collected specimens to determine the shape and colour of the lobes, the presence or absence of soredia and isidia and the size and septation of the spores within the fruit bodies. (However, it is important to avoid collecting excessive numbers of specimens for this purpose in the interests of conservation: see p. 59). There are also simple chemical tests that help to distinguish certain species. These tests can be performed at home or in the school laboratory and involve applying a tiny drop of test chemical (potassium hydroxide, sodium hypochlorite or paraphenylenediamine) using the necessary safety precautions. A colour change, usually to red or yellow, is then looked for. With the help of such observations, identification keys can be used to identify each lichen specimen. Keys are found in lichen floras which are published for Britain (for example Dobson, 1992) and for most parts of the world. Such books are listed in a recent compendium (Hawksworth & Ahti, 1990) and they can usually be obtained from public or university libraries. The lichens most often used for pollution studies in Britain and Ireland can be named with the help of plates 1–4 and the identification key on pp. 43–52. A very good way to become familiar with lichens is to attend one of the field courses on this group of plants which are run annually by the Field Studies Council. Field meetings and workshops are also run by the various lichen societies (see p. 74) and these too enable students and amateurs to improve their ability to identify lichens.

3 Sulphur dioxide and acid rain

Sulphur dioxide (SO_2) is a major component of urban and industrial air pollution which has been shown in field and laboratory studies to be very toxic to lichens and can kill sensitive species. It is a very soluble gas which can dissolve in rainwater or in moisture within the cell walls of wet lichen thalli. The cell membrane provides little barrier against the passage of sulphur dioxide, in its dissolved form (sulphurous acid), into the cell. At low pH, dissolved sulphur dioxide remains undissociated, but at higher pH levels, bisulphite or sulphite ions are formed; all forms are highly reactive. Sulphurous acid itself has weak oxidising properties while sulphite ions have quite strong reducing properties. Because dissolved sulphur dioxide is so reactive, it can disrupt a wide variety of metabolic processes in lichens including nitrogen fixation, photosynthesis and respiration.

As mentioned earlier, lichens are sensitive to sulphur dioxide because they have no stomata or protective cuticle (Richardson, 1988). In flowering plants, this pollutant is predominantly metabolised in the chloroplasts of the leaf cells, and the same appears to be true for the algae within lichens (Lange and others, 1989). Thus, lichens transplanted into sulphur dioxide-polluted areas, or fumigated with sulphur dioxide, exhibit deformation of chloroplast membranes and swelling of the mitochondria as symptoms of damage. However, recent studies in Switzerland on thalli of *Parmelia sulcata* growing naturally in various pollution zones around a city have shown no significant differences in the overall protein content, dark respiration or photosynthetic rates between urban and suburban habitats. In the urban lichen thalli, growth was seven times slower and, in addition, 15 times less assimilate was released by the photosynthetic partner to the fungus. It would therefore appear that the mechanisms involved in the release and transfer of carbohydrates from alga to fungus are even more vulnerable to the effects of SO_2 exposure than the chloroplasts. Surprisingly, samples from the more urban areas often had a higher chlorophyll content, perhaps because of the stimulatory effect of nitrogen oxides from traffic emissions on chlorophyll synthesis (von Arb & Brunold, 1989).

3.1 Effects of SO_2 on lichens

It is remarkably difficult to assess the viability, or the metabolic activity, of the two partners in lichens collected from polluted areas. Generally, the facilities required are beyond the scope of most school or college laboratories. One problem in measuring photosynthesis or respiration is that the lichen must be kept moist but not saturated. This is not easy because lichens gain or lose water rapidly and the water content greatly affects the rate of these processes. Furthermore, in determining the chlorophyll content of lichens, standard techniques such as acetone extraction are not suitable. Dimethyl sulphoxide is used because lichens contain substances that quickly convert the released chlorophyll to brown phaeophytin. When comparing the chlorophyll content of two lichen samples, one has to decide whether to present the results per unit dry weight or per unit surface area. The latter is difficult to determine especially in shrubby lichens. The dilemma arises because, unlike flowering plants, lichens are largely composed of fungal tissue which may form a thicker or thinner layer beneath the photosynthetic partner depending on the viability of the sample (whether it is just surviving or growing well). Using a sharp razor blade, thin sections of lichens from sites of different pollution levels may be cut and examined under a microscope. The proportion of plasmolysed cells can be estimated, but a fluorescence microscope is really needed to assess algal viability with any ease. Healthy algal cells within lichens fluoresce bright red in ultraviolet light while dying or dead cells exhibit an orange or white fluorescence (Holopainen & Kauppi, 1989). Recent studies have shown that it is possible to examine the spectral responses of lichens from polluted and unpolluted sites without cutting sections by using infra-red colour photography. After digitising and processing the data by computer, the effects of the pollution can be assessed (Gouaux & Vincent, 1990). The ultimate technique, transmission electron microscopy of extremely thin lichen sections, will reveal perturbations of the ultrastructure of cell organelles such as mitochondria or chloroplasts.

A technique within the capability of many schools or colleges involves measuring electrolyte leakage from lichen samples collected at different sites (p. 61). Changes in the conductivity of the water in which the lichens have been soaked provide a measure of the rate of potassium leakage from the cells of the lichen which in turn may reflect membrane damage due to air pollution (Alebic-Juretic & Arko-Pijevac, 1989). Using this technique, electrolyte leakage in *Rhizoplaca melanophthalma* was shown to decline significantly with distance from the coal fired Navajo electricity generating station near Page, Arizona. Chlorophyll degradation varied in parallel with electrolyte leakage. Arizona is a semi-desert area where the larger leafy and shrubby lichens are not generally abundant. This study shows that the smaller lichens can be valuable pollution indicators in such habitats (Belnap & Harper, 1990).

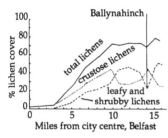

Fig. 2. The percentage cover of
lichens on deciduous tree
trunks at increasing distance
from Belfast, Ireland, along the
main road which runs south to
Newcastle (from Fenton, 1960).

3.2 SO₂ and lichen distribution

The diversity, abundance and cover of lichens on
tree trunks normally increase with distance from an
urban or industrial centre to an adjacent rural area. This
results from the different sensitivities of various species
to sulphur dioxide, with only a few being pollution-
tolerant (fig. 2). From empirical observations around
urban centres, coupled with readings from mechanical
monitoring stations, it has been possible to determine the
relative sensitivities of various lichen species to sulphur
dioxide. For example, the 'Hawskworth and Rose scale',
which distinguishes ten zones, relates the occurrence of
particular lichen species to mean winter sulphur dioxide
levels (Hawskworth & Rose, 1976) (table 1). The scale has

Table 1. *The 'Hawksworth and Rose' zone scale for the estimation of mean winter sulphur
dioxide levels in England and Wales using lichens growing on acidic and not nutrient-
enriched tree bark*

Zone	Moderately acid bark	Mean winter SO₂ ($\mu g/m^3$)
1	Algae only, e.g. *Desmococcus viridis*, present but confined to base.	> 170
2	Algae extends up the trunk; *Lecanora conizaeoides* present but confined to the bases.	About 150
3	*Lecanora conizaeoides* extends up the trunk; *Lepraria incana* s.l. becomes frequent on the bases.	About 125
4	*Hypogymnia physodes* and/or *Parmelia saxatilis* or *P. sulcata* appear on the bases, do not extend up the trunks. *Hypocenomyce scalaris*, *Lecanora expallens* and *Chaenotheca ferruginea* often present.	About 70
5	*Hypogymnia physodes* or *Parmelia saxatilis* extends up the trunk to 2.5 m or more; *P. glabratula*, *P. subrudecta*, *Foraminella ambigua* and *Lecanora chlorotera* appear; *Calicium viride*, *Chrysothrix candelaris* and *Pertusaria amara* may occur; *Ramalina farinacea* and *Evernia prunastri* if present largely confined to the bases; *Platismatia glauca* may be present on horizontal branches.	About 60
6	*Parmelia caperata* present at least on the base; rich in species of *Pertusaria* (e.g. *P. albescens*, *P. hymenea*) and *Parmelia* (e.g. *P. revoluta* (except in NE), *P. tiliacea*, *P. exasperatula* (in N)); *Graphis elegans* appearing; *Pseudevernia furfuracea* and *Bryoria fuscescens* present in upland areas.	About 50
7	*Parmelia caperata*, *P. revoluta* (except in NE), *P. tiliacea*, *P. exasperatula* (in N) extend up the trunk; *Usnea subfloridana*, *Pertusaria hemisphaerica*, *Rinodina roboris* (in S) and *Arthonia impolita* (in E) appear.	About 40
8	*Usnea ceratina*, *Parmelia perlata*, or *P. reticulata* (S and W) appear; *Rinodina roboris* extends up the trunk (in S); *Normandina pulchella* and *U. rubicunda* (in S) usually present.	About 35
9	*Lobaria pulmonaria*, *L. amplissima*, *Pachyphiale cornea*, *Dimerella lutea*, or *Usnea florida* present; if these absent, crustose flora well developed with often more than 25 species on larger well-lit trees.	Under 30
10	*L. amplissima*, *L. scrobiculata*, *Sticta limbata*, *Pannaria* species, *Usnea articulata*, *U. filipendula*, or *Teloschistes flavicans* present to locally abundant.	'Pure'

From Hawksworth & Rose, 1976 (revised).

☐ Zones 4-5   ▦ Zones 2-3   ■ Zone 1

Fig. 3. A simplified map of the five air pollution zones defined in Dublin, Ireland with the help of lichen data collected by school pupils. The mean winter sulphur dioxide levels in the zones were estimated to be: zone 1, >50; zones 2-3, 20-50; zones 4-5, <20 µg m⁻³ (after Ni Lamhna and others, 1988).

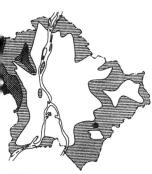

☐ desert zone
▤ inner struggle zone
▦ outer struggle zone
■ normal zone

Fig. 4. The air pollution zones of Budapest based on the distribution of lichens on tree trunks. Only in zone 4 was a relatively rich lichen flora found. In zones 2 and 3 pollution-tolerant species occur (from Farkas and others, 1985).

been widely used in Europe, sometimes in a modified form. Thus, five air quality zones were established in a study of the lichens of Dublin, Ireland, which involved surveying 2,215 trees in 257 1-km squares with the help of senior schoolchildren (Ni Lamhna and others, 1988) (fig. 3). There was an inner 'lichen desert' (zone 1) where no lichens were found. Outside this, four zones were defined in terms of the innermost occurrence of particular lichens: zone 2 by *Lecanora conizaeoides* and *Lepraria incana*; zone 3 by *Parmelia saxatilis, Parmelia sulcata* and *Hypogymnia physodes*; zone 4 by *Evernia prunastri, Parmelia caperata* and *Platismatia glauca*; and zone 5 by species of *Usnea* and *Ramalina*.

In studies that use the Hawksworth and Rose scale, it is often stated that, for example, *Lecanora conizaeoides* can withstand about 150 µg m⁻³, *Parmelia caperata* up to about 40 µg m⁻³ and *Usnea* up to 30 µg m⁻³ of SO_2 (Lerond, 1978). Data from recent studies in Ireland show that air quality zones established on the basis of lichen distribution correlate with much lower SO_2 levels. Thus in Ireland the limit for growth of *Lecanora conizaeoides* appears to be about 50 µg m⁻³ while that for *Parmelia caperata* is 30 µg m⁻³ and that for *Ramalina fastigiata* about 10 µg m⁻³. Possibly, the larger number of rain events in Ireland may keep the lichens moister and metabolically active for a longer period and hence they may absorb more sulphur dioxide and be more sensitive. Alternatively, past records of sulphur dioxide may have been calculated from mechanical monitors that measured total acidity. The Hawksworth and Rose scale therefore provides a usable indication of the relative sensitivity of individual lichens but the tolerated levels of sulphur dioxide pollution may be different, and should where possible be determined, at each study site. Even if absolute thresholds for damage are unknown, the technique is still useful for establishing the pattern of the pollution zones around a city or industrial area.

Outside Western Europe, other scales have been used. For example, in Budapest, a conurbation of two million people, four air pollution zones have been distinguished. Zone 1, which is devoid of lichens, corresponds approximately to the boundary of the built-up area of the two cities of Buda and Pest, lying either side of the river Danube. This zone has a mean winter sulphur dioxide level in excess of 300 µg m⁻³ (fig. 4). Outside this are two struggle zones where a series of resistant crustose species grow on the trees. The lichens in zone 2 include *Scoliciosporum chlorococcum, Lecanora conizaeoides, Buellia punctata, Lecanora piniperda* and *L. dispersa*. In zone 3, which is restricted to the more elevated areas of Pest and the peripheral areas of Buda, additional crustose lichens are found. Zone 4 is characterised by a more or less normal lichen flora on the trees, including leafy species like *Parmelia glabratula*. This zone is limited to a small area of Buda (Farkas and others, 1985).

The kind of distribution mapping described above has the disadvantage that a species must have disappeared before an effect is registered. Changes in pollution level can be monitored earlier by detailed notes or photographic records of any dying or discoloured lichen lobes at each site. These can be supplemented by records of the largest-sized thallus of individual species, such as *Evernia prunastri*. This was done around Newcastle upon Tyne in the UK (Gilbert, 1968). The most detailed way to follow gradual changes in lichen communities caused by air pollution is to record the occurrence of all species at each site together with a measure of frequency and/or percentage cover. This clearly requires greater taxonomic expertise and is also more time consuming, but it makes it possible to calculate Indices of Atmospheric Purity (IAPs) (p. 59). Maps can then be drawn with lines joining sites where the indices are numerically similar. The use of IAPs was first developed in 1964 by De Sloover in Canada and later employed to map air quality over the city of Montreal (LeBlanc & De Sloover, 1970). A recent study (Herben & Liska, 1986) suggests that IAPs can even be calculated without recording the percentage cover of each recorded species, as was thought necessary in the past.

The reliability of IAP studies increases with the proportion of pollution-sensitive species recorded in a survey. The lower the proportion of such species, the more trees should be surveyed at each site. A low total number of species for a survey cannot, however, be compensated for by surveying more trees. The IAP method is not therefore recommended where the whole study area falls into an area with an impoverished epiphyte flora (Herben & Liska, 1986). Recently, computer assisted techniques for analysing lichen distribution data have been used to produce three-dimensional IAP maps, as in the study of air quality around La Spezia in northern Italy (Will-Wolf, 1988; Nimis and others, 1990) (fig. 5).

It is important to determine if there have been any marked changes in sulphur dioxide levels in the area being studied, whether it be a city or an industrial complex. If levels are static or rising, distribution or IAP studies are likely to be useful predictors of current air quality. If, however, sulphur dioxide levels have fallen rapidly within the last five or ten years (following the implementation of clean air legislation, construction of tall emission stacks, factory closures or the introduction of new 'clean' energy sources like natural gas), then the distribution of recorded lichens may not relate well to measured pollution levels (see below).

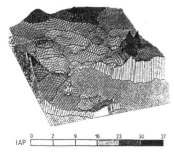

IAP   0   2   9   16   23   30   37

Fig. 5. A three-dimensional IAP map of the area surrounding the town of La Spezia, Italy, where there is a large coal-fired electricity generating plant. The dark elevated areas with high IAP values are those with pure air and a rich lichen flora (from Nimis and others, 1990).

3.3 Consequences of falling SO$_2$ levels

In recent years, several lichen species in Britain have shown a remarkable change in distribution as a

result of the falling ground-level concentrations of
sulphur dioxide. For example, in northwest London, the
mean winter sulphur dioxide level fell from around 130
μg m^{-3} in 1980 to between 55 and 29 μg m^{-3} in 1988. A
group of lichens assigned to zones 4 and 5 on the
Hawksworth and Rose scale failed to colonise the trees
although more pollution-sensitive species typical of
zones 6 and 7 did so. Thus some pollution-sensitive
species appeared able to recolonise areas more quickly
(perhaps gaining a foothold via vegetative propagules
like soredia or isidia) than the tolerant species. Therefore,
recolonisation following a fall in sulphur dioxide levels
need not necessarily follow an orderly sequence with the
more pollution-tolerant species invading first
(Hawksworth & McManus, 1989).

The spread of some species such as *Xanthoria
polycarpa* into northwest London was thought to be
promoted by nutrient-rich dust. Further evidence for the
importance of agrochemicals and other nutrient enriched
dusts in enhancing changes in urban lichen floras has
come from a study of Industrial Teesside in northern
Britain. There, the trunks of trees that are impregnated
with these dusts exhibit a flora that includes species more
commonly found on calcareous rocks, such as *Xanthoria*
species, *Lecanora dispersa*, *Phaeophyscia orbicularis*, *Rinodina
gennarii* and *Scoliciosporum umbrinum* (Seaward, 1990).

An interesting example of a lichen genus which
has shown a dramatic change in its distribution in Britain
is *Usnea*. This lichen was widespread in Britain until
about AD 1800 but it then disappeared from an area
covering some 70,000 km^2 as a result of increased air
pollution, especially sulphur dioxide (Seaward, 1987).
However, as levels of this pollutant have fallen, certain
species of *Usnea*, particularly *Usnea subfloridana*, have re-
established on trees such as ash and willow (fig. 6).

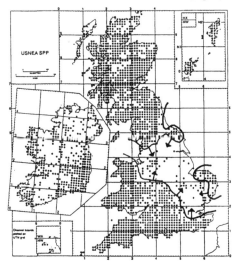

Fig. 6. The changing
distribution of *Usnea* in Britain
and Ireland following the
implementation of the Clean
Air Acts. The solid lines define
the areas from which this lichen
disappeared during the period
1800–1970. The arrows indicate
the recolonisation advance. The
dots are new records of *Usnea*
since 1970.
(Diagram supplied by M.R.D.
Seaward based on the data
bank of the British Lichen
Society Mapping Scheme at
Bradford University.)

USNEA SPP

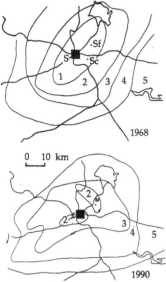

0   10 km

Fig. 7. Air quality zones in 1968
and 1990, based on IAPs
around the city of Sudbury,
Ontario. The change resulted
from a reduction of SO$_2$
emissions and the construction
of the world's tallest chimney.
The IAPs were calculated from
a study of lichens on balsam
poplar trees. S = Copper Cliff
smelter; Sf = Falconbridge
smelter and Sc = Coniston
smelter (now closed) (adapted
from Pappin & Beckett,
unpublished data 1991).

These trees have a bark with a higher pH than most other
deciduous trees, and it is better able to buffer the
substratum against the effects of acidic pollution.
Furthermore, the bark of ash and willow is rather softer
with a higher moisture retaining capacity. As a
consequence, lichens colonise such trees first in pollution-
stressed situations. The importance of bark acidity has
been revealed by an annual survey of the lichens of free
standing oaks along a 70 km transect out from the centre
of London. In spite of falling air pollution levels there was
little evidence for recolonisation. It is suspected that the
main reason for this is that oaks in the London area have
a highly acidic bark (pH 2.9–4.0) in comparison with other
tree species (Bates and others, 1990).

In Canada, the city of Sudbury, Ontario, is located
near one of the world's richest nickel deposits. The smelters
there have a production capacity of some two million
kilograms of nickel per year. As a result of the construction
of the world's tallest chimney (380 m), and an increase in the
amount of sulphuric acid produced as a by-product, the
amount of gaseous sulphur dioxide released in the Sudbury
area has fallen from over 2.5 million tons in 1960 to less
than a million tons in 1990. As a result there has been a
sharp fall in the ground-level sulphur dioxide levels and an
improvement in the diversity of lichens on balsam poplar
trees. The colonisation by lichens which reflects the
changing air quality is evident from a comparison of IAP
zones calculated from two sets of lichen data collected in
1968 and 1990 (Pappin & Beckett, unpublished data, 1991)
(fig. 7).

Changes are not limited to lichens on trees. A group
of pollution-tolerant crustose lichens including *Lecanora
muralis* has progressively invaded roofs and stonework in
what used to be lichen deserts of industrial cities throughout
England, such as the West Yorkshire conurbation, which
includes Bradford, Halifax, Huddersfield, Leeds and
Wakefield (Henderson-Sellers & Seaward, 1979) (fig. 8).

Fig. 8. The invasion of the West
Yorkshire, UK conurbation by
the lichen *Lecanora muralis*
between 1969 and 1990. The
black 1-km squares indicate re-
invasion by the lichen since the
implementation of the Clean
Air Acts of 1956 and 1968 (from
Henderson-Sellers & Seaward,
1979 and data supplied by
M.R.D. Seaward).

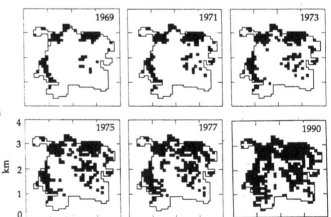

In addition, the foliose lichen *Hypogymnia physodes* has recently re-established at numerous sites within this conurbation (Seaward, 1989). These changing patterns have been revealed by detailed surveys of the same area at yearly intervals. Information on lichen distribution is stored on a computer at Bradford' University, the centre for the British Lichen Society Mapping Scheme, to which members submit lichen records. Changes in the distribution of particular lichen species in Britain and Ireland are monitored as part of the national mapping programme undertaken by the British Lichen Society using this database. Other types of information can also be extracted using various software packages.

The studies mentioned above have shown that lichens can re-invade an area within 5–10 years following a fall in pollution levels. However, such improvements in the lichen flora are less common than examples of progressive loss of lichens as a result of pollution associated with the spread of urbanisation and industrialisation (Hawksworth, 1990). This decline in the abundance and richness of the lichen flora has been documented in many countries by examining the present day lichen flora of a particular area and comparing the results with lists recorded in the past, or compiled from the study of old specimens in herbaria. For example, Arthur Mayfield, a headmaster of Mendelsham School in Suffolk, collected lichens in his area between 1912 and 1921. A re-survey in 1972–73 revealed that of 129 lichen species originally recorded, 62 could no longer be found. Air pollution and the effects of agricultural chemicals were considered to be the most important factors causing the change (Coppins & Lambley, 1974). Using a similar approach, the lichen floras of various national parks and recreation areas in the USA have been examined. For example, fewer than 20% of the 110 lichen species reported for the Indiana Dunes National Lakeshore area in 1896 could be found in 1985. Similarly, only 31 of the 151 species recorded in the Cuyahoga Valley National Recreation Area, Ohio between 1895 and 1917 could be found. In both cases, increased air pollution was mainly responsible for the observed changes. A survey was also undertaken in the Big Bend National Park, Texas following a report of decreased visibility due to haze. In contrast to the previous studies, there was no significant air pollution effect. There had been no loss of lichen species during a ten year period and there was little evidence to substantiate the claimed decline in visibility (Wetmore, 1989; O'Leary, 1988).

3.4 Acid rain

Acid rain is formed when sulphur oxides and nitrogen oxides originating from high temperature combustion sources such as electricity power generating stations are released from high chimney stacks so that the gases have a fairly long residence time in the atmosphere. As a result, sulphur dioxide is oxidised and falls as sulphuric acid, while nitric acid is formed from the

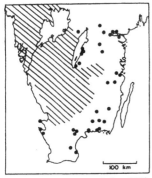

before 1950

1986–8

Fig. 9. The distribution of *Lobaria scrobiculata* in southern Sweden. The upper diagram is the known distribution before 1950 which includes scattered occurrences in the east and south as well as over 300 localities in the hatched area. The lower diagram shows the present distribution of the species (adapted from Hallinback, 1989).

nitrogen oxides. Acid rain tends to affect areas at some distance from the source, often being carried by prevailing winds to other countries. The main impact is to acidify the environment on which the acid rain falls, causing leaching of important nutrients or changes in the buffering capacity of bark or soil.

Epiphytic lichens, especially those containing cyanobacteria, seem to be particularly sensitive to the effects of acid rain. This has been well documented for *Lobaria scrobiculata* in Sweden, where the species was formerly recorded from more than 300 localities. It has now disappeared from all sites in the south and east of the country and has become rare elsewhere (fig. 9). Indeed, the presence of this lichen was recently confirmed at just two of 50 thoroughly investigated old sites (Hallinback, 1989). It seems that even weak acid rain from sources outside the country (transboundary sources) can progressively overcome the buffering capacity of tree bark to the point where the surface pH of the bark falls significantly (Nieboer and others, 1984). Such change may prevent the growth of cyanobacteria or of the *Lobaria* propagules. In an attempt to bypass this sensitive stage, a programme of transplanting mature thalli from an old unsound tree to young trees was initiated in Lowther Park, in the English Lake District. Ten of the fourteen transplanted *Lobaria amplissima* continued to thrive after ten years (Gilbert, 1990). *Lobaria* grows in ancient woodlands in areas remote from urban/industrial centres. Under the influence of acid rain, *Lobaria* is unable to thrive on the bark of a wide range of trees and has become restricted to those, like ash, which have bark with a higher pH and better buffering capacity. By recording the growth of *Lobaria* thalli in permanent quadrats over a long period using photography, it will be possible to assess the impact of acid rain even in remote regions.

Experimental studies on the effects of acid rain on lichens have so far been restricted to *Cladonia* species, which contain the green alga *Trebouxia* rather than cyanobacteria. Treating *Cladonia stellaris* in the field with simulated acid rain below pH 3.5 leads to thallus discoloration and a very gradual impairment of growth (Lechowicz, 1987). At higher pH, there is no discernible effect. Indeed, nitrate ions in simulated acid rain can cause an enhancement of growth which is greater in *C. rangiferina* than in *C. stellaris*. *C. stellaris* has been shown to be particularly pollution-sensitive (Scott and others, 1989). In acid rain with a low pH, the sulphate ions can depress growth and the molar ratio of sulphate to nitrate may therefore be important. The ratio of these two ions in the acid rain may therefore determine whether the overall effect will be harmful or beneficial for nutrient-poor lichen communities on peatlands or heathlands. In *C. stellaris*, simulated acid rain applied in laboratory experiments (pH <3.5 for two months) resulted in numerous plasmolysed and dead algal cells and damage to chlorophyll and photosynthetic systems (Roy-Arcand and others, 1989).

3.5 Sulphur accumulation

Lichens accumulate sulphur as a result of metabolising dissolved sulphur dioxide or taking up sulphate ions from acid rain or wind-blown sea spray. Lichen samples collected close to industrial complexes emitting sulphur dioxide have a much higher sulphur content than samples collected further away. *Cladonia mitis* collected ten miles from the Copper Cliff Nickel Smelter, Sudbury, Ontario, Canada had over 1000 µg g⁻¹ dry weight of this element, more than twice the local background level (Nieboer & Richardson, 1981). Similarly, a study of sulphur accumulation by *Hypogymnia physodes* over the whole of Finland showed that levels varied from a little over 400 µg g⁻¹ in the extreme north to 1830 µg g⁻¹ in the south of the country where there are industrial sources (Takala and others, 1985). Sulphur levels in the lichens correlated closely with wet and dry sulphate deposition. Figures for *Cladonia rangiferina* from Canada are somewhat lower but show the same trends (Zakshek & Puckett, 1986).

One important question is whether lichens take up sulphur by dry deposition, from rainwater, from leachates of sulphur-rich compounds from trees or from underlying soil or rocks. Recently isotope abundance studies have provided useful information. In Alberta, Canada, more than 1200 tons of sulphur, mostly as sulphur dioxide, are discharged per day as a by-product of natural gas production. The sulphur from these 'sour-gas wells' is enriched with the heavy isotope of sulphur, ^{34}S. As shown in figure 10, ^{34}S enrichment occurs in lichens and in the air around the gas wells but only to a small extent in pine needles from the region. Thus, lichens absorb a much higher proportion of their sulphur content from the air than do coniferous trees, for which soil is the primary source for this element (Krouse, 1977). Furthermore, the high ^{34}S enrichment in lichens implies direct absorption from the air or rain rather than from tree leachates (Takala and others, 1991). A final interesting observation was that lichens around the gas wells were able to release a proportion of their accumulated sulphur in the form of the lighter ^{32}S isotope. Thus lichens, like some flowering plants, can fractionate sulphur isotopes by emitting volatile sulphur compounds like hydrogen sulphide (Case & Krouse, 1980). The release of this gas is one way by which lichens can prevent a build-up of excess sulphur dioxide in the cells. Alternatively, lichens can metabolise sulphur dioxide via an oxidation process but this leads to sulphate accumulation and increases the danger of cell acidification (Lange and others, 1989).

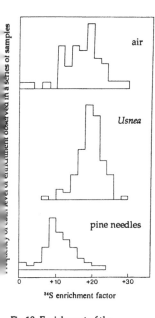

Fig 10. Enrichment of the ^{34}sulphur isotope in air, the lichen, *Usnea*, and pine needles collected near a sour-gas producing area of Ram River, Alberta, Canada. The enrichment was measured relative to the $^{34}S/^{32}S$ abundance ratio in meteoritic troilite (from Krouse, 1977).

4 Ozone and nitrogen compounds

Ozone (O_3) has become an important pollutant since the use of motor vehicles became widespread. Ozone is formed by an interaction between nitrogen oxides and various unburned hydrocarbons from vehicle exhausts; the reaction is mediated by sunlight. Ozone is thus an important pollutant in summertime in parts of the USA, particularly California, as well as in continental areas of Europe. In higher plants, the double bonds of fatty acids in the cell membrane are the main target of attack, with the chloroplasts being less affected (Lange and others, 1989). Ozone can be shown to damage lichens in laboratory experiments. It is more difficult to demonstrate that ozone is harmful to lichens in their natural habitat, partly because ozone is generally formed under dry sunny conditions when lichens are metabolically inactive. The Los Angeles region is one area where ozone and nitrogen oxides are particularly important components of the ambient air pollution. Ozone has been blamed for the marked decline of lichens on conifer and oak trees in the inland mountains, and oak trees in the coastal areas. At sites in the San Bernadino Mountains, the lichen cover was inversely related to the estimated oxidant air pollution doses (Sigal & Nash, 1983).

Ramalina menziesii, a large shrubby lichen, grows on trees of the coastal region. It occurs from southern Alaska to central Baja California, but has disappeared from the Los Angeles area of California where it was common early this century. To monitor the effects of pollution in the area, samples of this lichen were transplanted to a polluted site 35 km northeast of the city centre on the southern slope of the San Gabriel Mountains, and to a control site where the lichen occurred naturally in the Cleveland National Forest, adjacent to Palomar Mountain, about 200 km southeast of Los Angeles. Transplantation was carried out in both summer and winter. Gases and other substances were absorbed by the lichens during the summer dry period and the resultant ions were leached when lichen samples were placed in distilled water for brief periods. The highest levels occurred in samples from the polluted site at the end of the summer period. Mean values up to 170, 60, 45, 25 and 2 μmol g^{-1} were found respectively for nitrate, ammonium, hydrogen, chloride, fluoride, and sulphate ions. This suggests that for lichens, nitrogen compounds are one harmful component of air pollution in the Los Angeles area (Boonpragob and others, 1989). Laboratory studies revealed that chlorophyll levels and photosynthesis declined in transplants left in the polluted site during summer but not in transplants exposed for the winter period. This, together with observations on trace elements, suggests that dry deposition is the major pathway of accumulation in arid-region lichens (Boonpragob & Nash, 1990, 1991). For lichens growing in most other habitats, wet deposition seems more important (see section 9.1).

The damaging effects of nitrogen compounds on lichens have also been noted in Europe, especially in The Netherlands and Denmark. The lichen vegetation was studied on oak trees along two transects from a nature reserve to an intensive cattle-rearing area where there were high ammonia emissions. The abundance of each lichen species was recorded on ten trees at each site on a scale of 1–6 (1 = extremely rare – a single thallus; 2 = very rare – several thalli but only on one tree; 3 = rare; 4 = occasional; 5 = frequent and 6 = a species that is dominant at a site). The results showed that typical acid bark lichens such as *Evernia prunastri, Lecanora conizaeoides* and *Lepraria incana* occurred in the reserve, whereas in the cattle-rearing area they were replaced by species of *Physcia, Xanthoria* and other genera typical of eutrophicated habitats like the seashore and bird-perching stones (De Bakker, 1989).

Since the early 1980s, ammonia emissions have been reported to injure populations of *Cladonia portentosa, C. arbuscula* and *C. ciliata* in nutrient-poor habitats on dunes and heathlands in The Netherlands and Denmark. The lichen mats developed black edges and in some places, like the coastal dunes in North Jutland, large coherent areas of dead lichens were seen. The deposition rate of nitrogen has been estimated by measuring the nitrogen content of these lichens in different parts of Europe. Nitrogen contents varied from about 4 mg g^{-1} dry weight in western Ireland, to about 7 mg g^{-1} in Denmark and over 10 mg g^{-1} in parts of The Netherlands. The levels in these ground-dwelling lichens largely reflected wet deposition of nitrogen compounds. Epiphytic lichens proved to be better monitors for estimating dry deposition of ammonia, ammonium, nitrate and nitrogen oxides (Sochting, 1987, 1990).

5 Fluorides

The release of fluorides to the atmosphere is associated with industrial processes such as aluminium smelting, brick firing, glass making, and the production of fertilisers and phosphorus. The damaging effects tend to be more localised than those from sulphur dioxide emissions. Fluorides are also released in large amounts by volcanic eruptions and these emissions can cause illness or death to animals. The most famous example occurred in Iceland in 1783, when a 15-mile-long fissure opened up in the south-central part of the country and erupted for eight months. Sheep were killed by the emissions and the people were so weakened that one third of the population died (McPhee, 1989). Using lichens, fluoride releases can be studied by recording damage symptoms, by assessing changes in community composition, and by measuring the concentration of fluorides in the surviving thalli.

5.1 Damage symptoms

In December 1970, a new aluminium smelter producing 100,000 tons of metal a year was commissioned near Holyhead in North Wales. This provided the opportunity for a detailed study of the effects of fluoride pollution on lichens. Within 6–12 weeks of the start-up, damage was seen in nearby lichens, and it increased over the following two years. Damage was greatest on the sides of trees facing the smelter and involved four types of symptom:

1. Bleaching of the thalli, particularly in *Parmelia caperata* and *Evernia prunastri*. This is the result of chlorophyll breakdown in the algal partner.

2. Development of a red coloration in *Parmelia saxatilis*, *Parmelia sulcata* and *Hypogymnia physodes*. This seems to be a stress reaction by those lichens containing salazinic or physodallic acids. It is possible that fluorides stimulate urease production in these lichens and the released ammonia reacts with the β-orcinol depsidones like salazinic acid to produce a red colour.

3. Blackening and death of the thalli usually following the above symptoms. Sometimes, as in *Ramalina siliquosa*, the tips of the thallus die first.

4. Weakening of the attachment of the thalli to the bark resulting in loosening or detachment of the lichens, and their disappearance from the plant community.

On trees within 1 km of the smelter, shrubby lichens were most affected, with less than 1% of the initial cover remaining after five years. Leafy lichens were affected a little later and to a lesser degree, with about 12% cover surviving after the same time period. Crustose lichens were the least changed and some species, such as the relatively pollution-tolerant *Lecanora expallens*, exhibited a marked increase in cover. Within the leafy lichen group, there was a sequence of increasing sensitivity: *Parmelia perlata*, *P. caperata*, and to a lesser extent, *P. subrudecta* were more tolerant than *Hypogymnia physodes*, *Parmelia saxatilis* or *P. sulcata* (Perkins & Millar, 1987a). The last three species contain salazinic acid and seem to be especially sensitive to fluoride exposure. It is interesting that the above sensitivity sequence is the reverse of that found in areas polluted by sulphur dioxide.

On rocks, lichens developed symptoms of fluoride damage more slowly but damage was progressive. By 1985, after some 15 years' operation of the aluminium smelter, lichens at about half of the 61 studied sites within 2 km of the smelter were affected (Perkins & Millar, 1987b). Again, crustose lichens showed considerable tolerance to fluorides but little growth occurred in thalli from the most polluted areas. However, as a result of decreasing emissions since 1978, the species growing on rocks have recovered to some extent. This has occurred by the regrowth of previously

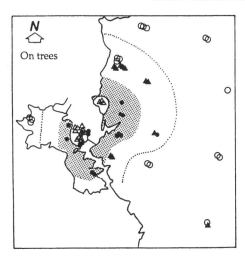
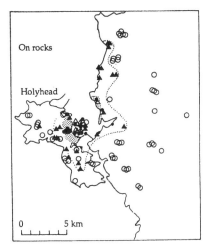

Fig. 11. A comparison of the extent of damage sustained by lichens on trees versus lichens on rock around the aluminium works (+) in northwest Anglesey, Wales from 1971 to 1985. Category 1 damage = 75% loss of cover (Δ); category 2 (● and shaded) = 40–75% loss; category 3 (▲) = 40% loss; category 4 (o) = no damage. Lines delineate areas of approximately equal damage (adapted from Perkins and Millar, 1987a, b).

damaged thalli rather than colonisation by new thalli (Perkins & Millar, 1987b). A comparison of the extent of damage on trees and rocks is shown in figure 11.

In the absence of baseline and follow-up studies such as the one in North Wales, the impact of fluoride emissions on lichen communities can be studied by carrying out lichen distribution surveys. This is done in a similar way to that described above for assessing the effects of SO₂ around urban areas. One such survey was carried out around an aluminium smelter in Arvida, Quebec, Canada. Indices of atmospheric purity were calculated and a map drawn which delimited six damage zones (fig. 12) (LeBlanc and others, 1972).

5 *Fluorides*

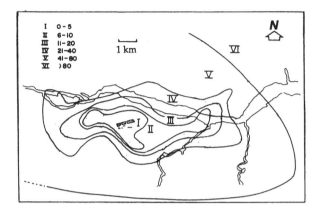

Fig. 12. Air quality zones around the aluminium works in Arvida, Quebec, Canada where volatile fluorides and gaseous hydrogen fluoride are released. The six zones are based on IAPs calculated from a study of lichens and mosses on balsam poplar trees (from LeBlanc and others, 1972).

At the ultrastructural level, studies on lichens growing near a ceramics factory and a fertiliser production facility in Finland showed that fluorides cause membranes within the chloroplasts of the algal cells to become swollen. The membranes then break down and small globules and crystalline material accumulate with concomitant mitochondrial degeneration (Holopainen, 1984a). Such symptoms have been duplicated in laboratory fumigation studies.

5.2 Fluoride accumulation

Fluoride levels in lichens are measured using a specific ion electrode after cleaning, drying and fusing with alkali. It is not easy, without experience, to make determinations accurately. Before the commissioning of the aluminium smelter at Holyhead, North Wales, the lichens contained less than 10 μg g^{-1} of fluoride but within six months enhanced fluoride levels were being detected. As production increased, levels over 600 μg g^{-1} were recorded in specimens collected within 1 km of the source. Such lichens soon died and disappeared. Although emissions have decreased, enhanced fluoride levels (sometimes over 50 μg g^{-1}) still occur in lichens, particularly during dry weather, at distances up to 25 km from the works. With the onset of rain, these levels return to near normal values within 15 weeks. Many leafy and shrubby lichens seem able to accumulate up to about 50 μg g^{-1} without being permanently damaged. *Parmelia loxodes* is more tolerant, showing little or no injury even after taking up about three times this amount of fluoride. The pattern of fluoride fallout is such that the fluoride concentrations in *Ramalina siliquosa* exceed 100 μg g^{-1} in samples collected near the smelter but this falls rather quickly to about 20 μg g^{-1} at 15 km and to very low levels at twice this distance (Perkins, 1980).

Fluorides are formed from one of a group of elements known as the halogens. Chlorides, which lichens readily accumulate, are formed from another halogen, chlorine. In Finland, *Hypogymnia physodes* and *Pseudevernia furfuracea* exhibited mean chloride levels of 1500 and 3000 μg g^{-1} dry weight respectively and nearly twice this level near the sea. There, wind-blown salt spray provides a source of chlorides. These two lichens synthesise chloratranorin (a chlorine-containing lichen compound) in their cortex, and may thereby gain protection against excess chloride levels (Takala and others, 1990). As lichens accumulate high chloride levels from wind-blown salt particles without damage, it seems likely that it is the hydrogen fluoride, rather than the particulate fluorides, that causes much of the damage to lichens around aluminium smelters.

5.3 Volcanoes

Volcanoes expel fluorides in gaseous form or as microscopic salt particles adsorbed onto volcanic ash. During an eruption, both particulates and gases are emitted in large volumes over a short period whereas during the quiescent phase fluorides form part of the gas plume. Monitoring this slow release of gaseous fluorides is difficult. While lichens cannot be used to give an absolute measure of the fluoride content of discharged gases, they can be used to determine the dispersion characteristics and long-term output. For example, Mount Etna in Sicily, which has a summit over 3,000 m above sea level, is one of the world's most active volcanoes. Samples of *Xanthoria parietina* and *Stereocaulon vesuvianum* were collected from 77 sites on the volcano in 1985, and from 56 sites in 1987. The samples showed high fluoride levels on the northeast and east slopes of the mountain. The highest values of 141 μg g^{-1} occurred at a distance of 5 km from the source of the plume in and around the Valle del Bove (fig. 13). The lichen data revealed that the principal factor controlling fluoride fallout was the prevailing wind but this was modified to a considerable degree by the shape of the mountain. The northeast rift zone forms a high ridge which diverts the plume, resulting in increased fluoride levels on the flanks, and funnels wind down into the Valle del Bove (Davies & Notcutt, 1988).

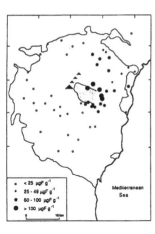

Fig. 13. The fluoride concentration per unit dry weight in lichen samples collected in 1987 at various locations on Mount Etna, Sicily, which is Europe's largest volcano (Davies & Notcutt, 1988).

Following volcanic eruptions, lichens can be used to help identify fluoride-contaminated areas. This is important because both animals and man can develop fluoridosis. Recent studies on La Palma in the Canary Islands have shown that lichens are valuable monitors of fluoride emissions, even from minor eruptions. They can also be used to study the release of toxic gases from lava as it cools. For example, at one location on La Palma the 1949 lava flow was less than 200 m wide and 1–2 m thick but it degassed enough to cause a marked increase in fluoride levels in lichens growing nearby. The lichen data revealed that fluorides may be released over a period of tens of years which contradicts the conventional view that degassing takes place rapidly as the lava cools (Davies & Notcutt, 1989).

6 Aromatic hydrocarbons

These potentially toxic pollutants are released from a variety of sources. Aromatic hydrocarbons and polyaromatic hydrocarbons (PAHs) include dioxins and furans (released during the incomplete burning of, for example, fossil fuels or rubbish) as well as PCBs (polychlorinated biphenyls). PCBs are used as stabilisers in the chemical industry and in electrical transformers. Spills of waste transformer fluid, or PCB releases from chemical processes, are especially serious because these compounds are potent carcinogens. PCBs are more volatile than PAHs and so occur in the atmosphere as gases rather than particulates. Chlorinated hydrocarbons (Lindane, DDT, HCH and HCB) include pesticides and compounds employed in the paint and plastics industry. The use of all these is restricted in most countries because of their long residence time in the environment and their potentially serious effects on human health.

Since about 1983, it has been realised that lichens accumulate aromatic hydrocarbons and can provide information on the degree of contamination of an environment. Determining the levels of these compounds in lichens involves freeze-drying the samples which are then extracted with organic solvents. The extracts are concentrated by evaporation, and separated into PCBs and pesticides by column chromatography. The various compounds are then quantified by gas chromatography, sometimes coupled with mass spectrometry. This type of sophisticated analysis was applied to samples of reindeer lichen *Cladonia rangiferina* collected between 1961 and 1972 from a remote area in northern Sweden. Over the ten year period, there was a progressive increase in the levels of PCBs (fig.14) in the lichens. The results suggested that aerial transport was the main route for the dispersal of chlorinated hydrocarbons to this isolated site. The mean residence time in the atmosphere was estimated to be 2–3 years, as there was a delay of this period between an increase (or maximum) in the production of a compound and an increase (or maximum) in the observed

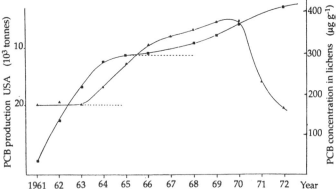

Fig. 14. The production of PCBs (▲) and the concentrations of these compounds found in carpets of the lichen *Cladonia rangiferina* (■) over the period 1961–72. The rise in PCB production in 1963 (dotted line) resulted in increased levels in the lichen 2 years later (dotted line) (from Villeneuve & Holm, 1984).

level in the lichens (Villeneuve & Holm, 1984). In winter, *Cladonia rangiferina* is a major food for reindeer which are eaten by the Lapp people. Studies of this food chain indicate that Lapps may ingest 12 μg per year of these compounds from deer meat, but fortunately this is a thousand times less than the allowable limit (Villeneuve & others, 1985).

People in urban areas are more at risk from pesticides and industrial contaminants. A recent report indicates that breast-fed babies in Canada ingest about 160 picograms of dioxins and furans for each kilogram of body weight. Mothers ingest these compounds from foods contaminated by numerous sources including wood preservatives, incinerator emissions and pulp mills (which use chlorine for bleaching the products). The amount the babies receive is 16 times more than the acceptable exposure standard. However, most babies are fed on breast milk for only a short period and the milk confers nutritional and immunological advantages that far outweigh the potential dangers from these levels of dioxins and furans (Kirkey, 1990).

Lichens have been used to assess contamination of the Antarctic by chlorinated compounds. HCH levels in samples from the Antarctic Peninsula were ten times lower than those in lichens from Sweden and Finland, and 100 times less than those from Italy. In contrast, HCB levels were similar at all sites. The Antarctic lichens are probably reflecting the mean level of the vapours of these compounds in the troposphere (Bacci and others, 1986).

Samples of *Usnea*, collected recently in southern France, contained concentrations of insecticides and the chlorinated hydrocarbon Toxaphene that generally increased with altitude. In contrast, levels of particular PCBs were irregularly distributed. The insecticides were probably derived from agriculture in the Rhône Valley, the higher elevation sites being more exposed to air masses moving from this area. The effectiveness of PCB and other aromatic hydrocarbon accumulation by lichens is illustrated by calculating concentration factors between lichens and the atmosphere. Generally for PCBs this factor is around 1×10^5, but for Arochlor it is up to three times higher. Although nothing is known about the mechanism and rates of absorption of chlorinated hydrocarbons in lichens, it is clear from the high concentration factors that lichens are excellent biomonitors for these compounds (Villeneuve and others, 1988). To date there have been no studies on the effects of chlorinated hydrocarbons on the physiology of lichens. However, other organic compounds such as the herbicide Ustinex (active ingredient diuron) have an inhibitory effect on lichen algae (Luhmann and others, 1989).

The chemical detection of specific aromatic hydrocarbons accumulated by lichens requires elaborate analytical techniques. In theory, pesticide concentrations might be estimated more simply using a carefully calibrated bioassay. For example, the mortality of fruitflies might be assessed following their exposure to lichen samples collected from contaminated areas. This possibility remains to be explored.

7 Metals

Lichens accumulate high levels of various metals and are thus excellent monitors of atmospheric fallout around smelters, industrial centres, urban areas, mines and road systems. To determine metal levels in lichens, samples are first cleaned of debris, and dead or discoloured portions are removed. The samples may be washed or simply dried before analysis. Extraneous surface dust can increase variability in the measured element levels, especially if samples are collected on different days or under various weather conditions. Washing not only removes this dust but also some of the more weakly bound elements such as potassium and calcium. Potassium can also leak from within the cells following rapid immersion in water. The effect of washing should always be checked on a few samples. If the expected contaminants are metals like copper, chromium and lead, then washing should have little effect and can be part of the experimental protocol. Once the samples have been dried, they are normally ground to a powder in a mortar using a pestle. The addition of a little liquid nitrogen makes lichens very brittle so that they can be crushed more easily. The metal content of the lichen powder is then measured by atomic absorption spectrometry, modern microprocessor-controlled polarography or inductively coupled plasma emission spectroscopy. Non-destructive methods like X-ray fluorescence spectrometry or neutron activation have the advantage that samples can be archived and re-analysed at a later date. Thus additional elements can be quantified in the orginal samples if later thought necessary. Furthermore, variations in year-to-year laboratory operation or the efficiency of the analytical instrument can be allowed for, and this is important when the aim is to establish absolute concentrations of elements in lichens or changes over time in the pattern of metal accumulation. For comparative studies around heavily contaminated areas, it is possible to use less sophisticated colorimetric methods which are widely available (see p. 62). However, their sensitivity would probably not be sufficient to detect levels above background at more distant sites.

Lichens accumulate metals by trapping insoluble particulates (usually oxides, sulphates and sulphides). They also take up dissolved metal ions onto their cell walls by ion exchange and this may be accompanied by a slower uptake of metals into the cells. Some features of these uptake systems are described below.

7.1 Uptake by ion exchange

Metal ions bind to sites on the cell walls of lichens. A large proportion of the uptake occurs within five minutes, and the process is complete within one hour of a lichen sample being placed in a solution of metal ions. It occurs as follows:

$$M^{2+} + \left[\begin{array}{c} \diagup COOH \\ \\ \diagdown COOH \end{array} \right. = \left. \begin{array}{c} \diagup COO \\ \\ \diagdown COO \end{array} \right] M + 2H^+$$

Where M^{2+} denotes the entering metal ion, $-COOH$ the protonated binding site on the lichen cell wall, $-COOM$ the metal bound to the cell wall and H^+ the released hydrogen ions. Metal ions bound to these ion exchange sites can be displaced by ions with greater binding affinity or by ions that have a lower affinity but are present at higher concentrations. The following affinity sequence has been established for the lichen *Umbilicaria muhlenbergii*: Cu> Pb > Zn > Ni > Mg > Sr > K. The metals are thought to bind to carboxylic acid groups which are probably part of proteins in the cell walls of the lichen (Richardson and others, 1985). Thus lichens act like ion exchange resins absorbing metal ions from rainwater and releasing hydrogen ions or metal ions of low binding affinity as uptake proceeds. This can be demonstrated experimentally by first loading a lichen with an ion like strontium and then placing the lichen in a solution of a second ion and measuring how much of the second metal ion binds to the lichen and how much strontium is released (table 2). Metals bind to the walls of both the fungi and algae (Xue and others, 1988; Tyler, 1989). Studies on the lichen *Umbilicaria muhlenbergii* show that higher concentrations of nickel ions bind to the layer of the lichen thallus which contains the algae. Overall, however,

Table 2. *The molar exchange ratios observed in lichen experiments with* Umbilicaria muhlenbergii *and* Cladonia rangiferina *showing that lichens behave like synthetic ion exchange resins*

Exchange	Metals taken up during treatment μ moles g^{-1}	Cation released by treatment μ moles g^{-1}	Sr:Metal molar ratio
Umbilicaria			
Ni^{2+} for Sr^{2+}	11.8	12.2	1:0.97
Sr^{2+} for Tl$^+$	4.1	8.6	1:2.1
Sr^{2+} for II$^+$	6.3	13.2	1:2.1
Cladonia			
Ni^{2+} for Sr^{2+}	28.8	29.7	1:1.0
Cu^{2+} for Sr^{2+}	31.9	28.5	1:1.1
Tl$^+$ for Sr^{2+}	13.8	6.8	1:2.0

From Richardson and Nieboer, 1981.

Table 3. *The uptake of nickel ions by different parts of the thallus of* Umbilicaria muhlenbergii. *The algal zone includes the upper protective layer (cortex); the medulla is the central storage area of the thallus and the plates are extensions of the lower surface which are peculiar to this lichen*

	Algal zone	Medulla	Plates
Percentage composition by weight of each fraction	15	46	39
Ni content as $\mu g\ g^{-1}$ dry weight	1314	1041	646
Ni content of each zone as a percentage of the total	21	52	27

From Richardson & Nieboer, 1983.

more is bound to the fungal component since the bulk of a lichen is composed of fungal tissue (table 3). In the case of lead uptake by *Ramalina duriaei*, ions appear to bind most to the two cortical regions as revealed by the red coloration in these outer zones when the lichen is subsequently treated with sodium rhodizonate (Garty & Theiss, 1990) (see p. 61).

When a lichen sample is incubated in a solution containing ions of various metals, it is found that particular metals show a preference for the type of binding group to which they attach. Metals can be grouped into three classes. The ions of Class A metals, which include Na, K, Ca, Mg and U, bind to carboxylic acid and other oxygen-containing groups. The ions of Class B metals will bind to these oxygen-containing groups but also attach very strongly to nitrogen and sulphur centres such as the amino acids and sulphydryl groups of enzymes. As a result, enzymes may cease to function. Thus Class B ions, such as Ag, Tl and Hg are very toxic to living organisms. The third class, termed Borderline metals, are intermediate in their binding preference. They bind to either type of binding group with an affinity depending on the metal and concentration. Many of these metals are essential micronutrients at low concentrations but are toxic when present in excess. In the latter situation, they may also inactivate enzymes and other biomolecules within cells and they generally exhibit the following order of increasing toxicity: Fe < Zn < Ni < Co < Cr < Cd < Cu < Pb (Nieboer & Richardson, 1980; Van Assche & Clijsters, 1990). This sequence, to a degree, reflects the increasing sulphur-binding or 'Class B' character of the metals.

7.2 Uptake into cells

The rapid uptake of metal ions by the process of ion exchange may also be accompanied by a slower uptake into the cells of the lichen (Brown, 1985). Some ions which are major nutrients, like potassium, are predominantly accumulated within the cell and bind only weakly and transiently to the cell wall. Others, like zinc, magnesium and calcium bind quite strongly to the cell wall but are also taken up into the cells. Studies on the free-living fungus *Penicillium notatum*, the original source of penicillin, suggest that essential metal ions are taken into fungal cells by energy-dependent uptake mechanisms and that each element has a specific uptake system (Starling & Ross, 1990). In short-term experiments lasting a few hours on the lichens *Peltigera* and *Cladonia*, less than 10% of the zinc ion supplied entered the cells. In contrast, around 50% of the total zinc content was located within the cell in samples collected from the natural habitat where there is a lower external concentration and longer time for cellular accumulation (Brown, 1985).

7.3 Trapping particulates

In most cases, metal-rich emissions from industrial processes are predominantly in the form of insoluble particulates. Lichens accumulate such particulates in the same way as they accumulate particulates derived from rocks and soil. Tiny particles become entrapped between the growing hyphae and often seem to be accumulated in the rather loose central medulla region of the thallus. The effectiveness of this process has been shown by studies of lichens growing around the city of Sendai, Japan. The metal content of lichens was compared with that found on filters which trapped urban particulates. It is evident from figure 15 (Saeki and others, 1977) that there is a linear correlation between the levels in the filters and that in the lichen suggesting that much of the metal accumulated by lichens is in particulate form. This is further confirmed by studies which attempted to displace metals bound to the cell walls of lichens using acid. Around the nickel smelter in Sudbury, Ontario, less than 20% of the accumulated iron could be displaced whereas over 60% of the zinc could.

The importance of particulate trapping has also been demonstrated in areas away from industrial activity. The levels of a range of metals, including iron and titanium, were measured in lichen samples. The iron/titanium ratios in the lichens were of the same magnitude as those generally found in rocks, indicating entrapment of rock particulates. Much of the accumulated iron, chromium, vanadium and nickel could also be assigned to particulates. This was deduced by comparing the observed levels of various elements with those predicted on the basis of the average element to titanium ratios found in the earth's crust. For potassium, a macronutrient, a large departure from the predicted value was found. The reason is that potassium, unlike the other elements, is taken up in soluble form and accumulated to high levels within cells. A marked departure from the expected ratio was also found in lichens from industrially contaminated areas (for example, around smelters or oil refineries) where either Fe or Ti was emitted (Richardson & Nieboer, 1981).

7.4 Damage symptoms

Lichens accumulate metals outside the cell membrane within the cell wall and between the cells

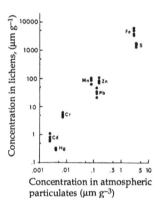

Fig. 15. The linear relationship between the metal content of lichens and that found in collected atmospheric particulates around the city of Sendai, Japan. The arithmetic means for the different elements in four species of *Parmelia* are plotted against the element content of particulates determined from the geometric means of samples collected from 1971 to 1974 (from Saeki and others, 1977).

(extracellularly). Often the total metal content of lichen samples can be high without apparent harm. Levels of some elements in excess of 5000 µg g^{-1} dry weight have been recorded (Nash, 1989; Tyler, 1989). It is the nature and the form of the metal that is important. This can be established by experimentation and by observation of lichens growing in particular habitats. For example, in Haiti, King Cristof (1807–20) built himself a French-style palace at sea level and an impregnable castle on a nearby mountain top (792 m above sea level). The French never re-invaded, with the result that the cannon and cannon balls are still in place. The iron cannon balls have become covered with a range of shrubby, leafy and crustose lichens, including *Chiodecton, Heterodermia, Ramalina, Parmelia* and *Teloschistes*. These are growing on the iron surface without apparent harm.

Zinc and particularly lead and copper are much more damaging to lichens than iron. The effects of zinc can be seen on tree trunks below the points at which zinc-coated barbed wire is attached, or on lichen-covered roofs in rural areas below the galvanised supports of television aerials. The effects of copper telephone wires can also be seen in such situations. Zinc or copper ions, leached from the metal, soon saturate the ion exchange sites on the cell walls and are taken up into the cells of the lichen where their excess is sufficient to kill the lichen. A similar situation can sometimes be seen where inscriptions on gravestones are inlaid with lead to make them more conspicuous. Lichens are able to colonise the whole gravestone with the exception of the area below the inscription; here the leached lead ions provide a highly toxic environment (Richardson, 1991).

Field observations have been supported by laboratory experiments. At the cellular level, excessive levels of Borderline or Class B metals bind to sulphydryl groups in enzymes and affect their function or structure. Furthermore, toxic ions can replace essential metal ions in a metalloprotein. Finally, in excess, metals like copper cause damage to the cell membrane, while zinc and cadmium accumulate in chloroplasts and can perturb their normal function (Van Assche & Clijsters, 1990). Some lichens secrete quite large amounts of oxalic acid and crystals can be found coating the fungal filaments. In such cases, excess metal ions may react with the oxalic acid to form insoluble metal oxalates (Jones, 1988). Such complexes with oxalic acid or other lichen compounds provide a metal tolerance mechanism and allow species like *Lecanora cascadensis* and *L. vinetorum* to grow, respectively, on copper-rich rocks or wood substrata sprayed with Cu-containing fungicides (Tyler, 1989). In other instances, the fungal strands (rhizinae) attaching lichens to their substratum may regulate the amounts of metal reaching the thallus from metal-contaminated soils as in *Peltigera* (Goyal & Seaward, 1982).

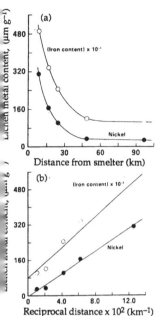

Fig. 16. The iron and nickel contents of *Stereocaulon* species (a) in relation to distance from the Copper Cliff smelter, Sudbury, Ontario, and (b) as a function of the reciprocal distance along the same transect. A plot of the type shown in (b) often yields a linear relationship which provides one way of evaluating background levels in the absence of pollution (from Richardson & Nieboer, 1981).

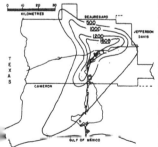

Fig. 17. The pattern of iron fallout based on the iron content ($\mu g\ g^{-1}$ dry weight) of *Parmelia praesorediosa* samples collected around the Calcasieu area of Louisiana, USA (from Thompson and others, 1987).

7.5 Accumulation around urban and industrial sites

Lichens are excellent monitors of the nature and extent of metal contamination around a wide range of industrial activities. Typically, high metal levels are found in lichen samples collected close to the emission source. These levels fall off rapidly at first and then more slowly as the distance from the source increases. This pattern was seen in samples of *Stereocaulon* species collected around Copper Cliff smelter, Sudbury, Ontario and analysed for nickel (fig. 16). The pattern of fallout may be affected by topography and prevailing wind direction. This was found on analysing *Parmelia praesorediosa* from the Calcasieu area of Louisiana, USA (Thompson and others, 1987). The pattern for iron fallout (fig. 17) was paralleled by that for chromium, zinc and mercury. The heavy concentration of industry in this area included two oil refineries, four chemical plants, a coke production facility and a chlor-alkali plant. Emissions from the last probably accounted for the elevated mercury levels. From 1982 to 1988 there was a dramatic fall in the levels of various elements in lichen samples collected from the studied sites (Walthier and others, 1990). This fall in elemental levels has resulted from a combination of more stringent pollution controls and a slowdown in industrial activity. By collecting data on metal levels in lichens annually, it was possible to calculate that the biological mean residence time was about four years for zinc but only half this for aluminium, iron and mercury. The greater residence time for zinc probably reflects the fact that it accumulates to significant levels within the cells of a lichen. As shown by these data, lichens can be expected to show a fall in elemental content after about two years following a reduction in emissions from an industrial source.

In many older urban or industrial areas in Western Europe, only the pollution-tolerant lichen *Lecanora conizaeoides* can be found in any quantity. This crustose lichen is more difficult to sample than leafy or shrubby lichens. In spite of this, *Lecanora conizaeoides* has been used to monitor metal fallout around a steel works and iron foundry at Frederiksvaerk, Denmark. Very high metal levels were found close to the foundry in dust resulting from the handling of raw materials. However, at this site the pattern of fallout was not as predicted from wind rose information. The zones established from the measured iron concentration in the lichens were clearly affected by the 50 m deep valley in which the foundry was located (Pilegaard, 1978).

Lecanora conizaeoides has also been used to monitor industrial emissions over a wide area of The Netherlands. By applying a statistical technique known as target transformation factor analysis to data on some 18 elements in the lichen samples, it was possible to identify various sources of pollution where emissions were spatially complex. Six factors were identified. Factors 1 and 2 showed aluminium, scandium and iron concentrations that characterise a contribution from soil to air particulate matter and reflect agricultural activity. Factor 3 combined a number of volatile elements including bromine from high temperature industrial activities located between Brussels and Antwerp. Factor 4 was characteristic for emissions from zinc smelters. Factor 5 contained cadmium and tungsten, probably from a factory making television tubes where phosphors were prepared and handled. Finally, factor 6, with mercury as a significant element, may represent areas where Hg-containing fungicides are employed in agriculture (De Bruin and others, 1986). Recently, the survey has been repeated on two occasions at an interval of five years using the foliose lichen *Parmelia sulcata*. As a result it was possible to follow changes in the pattern of metal pollution over the country (Sloof & Wolterbeek, 1991).

Highways are linear sources of pollution. Until the recent introduction of lead-free petrol, there was considerable effort made to monitor the fallout of lead in relation to distance from the road. The source of the lead is the antiknock compound lead tetraethyl which has been added to petrol since the 1940s in increasing amounts. Thus, along a busy four-lane highway in southern Finland, there was a statistically significant fall in lead levels in *Hypogymnia physodes* from about 175 µg g^{-1} dry wt at 20 m to around 75 µg g^{-1} at 100 m from the road. Quite high levels continued to occur up to the most distant sites at 200 m (Laarksovirta, 1976). It is interesting that samples of various *Parmelia* species collected 300 m from a highway in central Italy exhibited similar lead levels. The older parts of the thalli had higher concentrations of metal, indicating that for monitoring purposes, the same proportion of marginal lobes and central thallus areas should be analysed in all samples (Bargagli and others, 1987a). In a another study of *Hypogymnia physodes* collected near a lightly trafficked road in western Bohemia, Czechoslovakia, lead levels fell from 50 to around 10 µg g^{-1} dry wt within 200 m (Kral and others, 1989) (fig. 18). Cadmium concentrations in the lichens showed a similar pattern but were at much lower levels. Lead emitted from vehicle exhausts is a particular hazard for young children, with potentially damaging effects on the central nervous system even in small amounts (Richardson, 1982). As a result, leaded petrol is being phased out around the world. A consequent reduction in contamination levels has been demonstrated with the help of lichens even in rural areas. Thus lead levels in *Parmelia baltimorensis* collected in the 1950s in the Shenandoah National Park, Virginia, USA, through which Skyline Drive runs, had levels close to 200 µg g^{-1} dry weight whereas thalli collected in 1985–6 had only about 70 µg g^{-1} of lead (Lawrey & Hale, 1988). The value of lichens as quantitative monitors of lead and other airborne metal pollution has been demonstrated following analysis of metal levels in sequential growth increments from thalli of *Parmelia balimorensis* (Schwartzman & others, 1987).

In urban areas lead accumulation by lichens occurs mainly as a result of the deposition of lead particles on the surface under dry conditions (dry deposition). In contrast, in

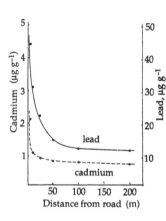

Fig. 18. The concentration of lead (——) and cadmium (----) in *Hypogymnia physodes* in relation to distance from a lightly used road in Czechoslovakia (from Kral and others, 1989).

rural areas some 80% of the lead taken up by lichens is dissolved in rainwater (wet deposition). At both sites a significant part of the dry-deposited lead can be solubilised following rain or dew. This is because hydrogen ions, released by the lichen, which has a relatively low natural pH, acidify the water films on and in a moist lichen. This acidity speeds up the solubilisation of the deposited lead particulates. The resultant lead ions can be accumulated on the cell walls of the lichen by ion exchange. From studies on lead uptake by *Parmelia baltimorensis* and of the lead content of this lichen collected near a highway in Maryland, USA, it has been possible to construct a computer model that reflects the drop in lead emissions from petrol over the period 1973–86. The application of this computer-assisted approach could help to identify trace metal 'signatures' in acid rain and give clues to the source of this pollutant (Schwartzman and others, 1991). If the source of particular episodes of acid rain can be identified, then the principle that a 'polluter pays for resultant damage' can be better enforced.

7.6 Accumulation near mines

The dust from mining activities is a potential hazard and originates from ore crushing, ventilation shafts, or the disposal of tailings (unwanted rock from which the metal-rich particles are removed by flotation or other means). This dust is most dangerous when the metal is either toxic or radioactive. Lichens may be used to define the zone of influence of mining activities on the surrounding area. This is done by collecting lichen samples which are then analysed for the appropriate metal. For example, *Parmelia sulcata* was collected from 50 sites around the former cinnabar mining area on Mount Amiata in Italy and the mercury levels were determined. Elevated levels were found on the eastern side of the mountain, reflecting the prevailing wind (fig. 19). It was interesting that the highest Hg levels in the lichen (about 8 μg g^{-1}) occurred near the mine ventilation systems (even though these were not functioning), and not near mine spoil heaps. Thus, the lichens had accumulated mercury both from gaseous (Hg vapour) emissions (from mine shafts, soil and vegetation) and by deposition from the atmosphere. Mercury levels on the western slope of the mountain were much lower, but still above that expected for rural areas. This may be explained by the presence of mineralised rock and some geothermal power stations (Bargagli and others, 1987b). The potential danger of mercury contamination from geothermal exploration boreholes and geothermal electricity generating plants has been realised recently. The Italian government passed a law in 1986 to regulate activities in geothermal areas as exploration is becoming intense (Baldi, 1988). The reason for this is that alternatives are being sought to fossil-fuel power

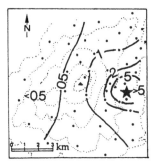

Fig. 19. The mercury concentration in *Parmelia sulcata* (as μg g^{-1} dry weight) growing near a former cinnabar mining area on Mount Amiata, Italy. The dots show collection sites and isometric lines delineate different contamination levels (from Bargagli and others, 1987b).

stations which emit carbon dioxide. This and other gases can cause global atmospheric warming, the so called 'greenhouse effect', which threatens world climates.

Uranium mining is another potentially hazardous activity. Uranium is a mildly radioactive element. The associated gas, radon, resulting from the radioactive decay of uranium has to be removed from mines via the ventilation shafts because it can induce lung cancer. The radium and thorium remaining in the tailings may also cause cancers. Studies around the Elliot Lake uranium mining area in Ontario, Canada showed uranium levels in *Cladonia rangiferina* up to about 12 µg g^{-1} dry weight in samples collected near the mines or ventilation shafts. This level fell rapidly as distance from the source increased. Mathematical estimates were made of the contaminated zone around Elliot Lake with the help of a regression model (p. 63). Using this model, the lichen data indicated that the polluted zone extended for up to 22 km around the mining area, whereas around individual mine-exhaust vents it was limited to about 400 m (Beckett and others, 1982). The entrapment of metal-rich rock particulates by lichens is thought to be the main mechanism resulting in elevated uranium levels.

Around most other types of mining or quarrying activities, lichens also accumulate metals by trapping particulates and without apparent harm because the metals are usually in rather insoluble form. The degree of particulate soil and rock accumulation can be evaluated by ashing lichen samples collected at increasing distances from the contamination source. The ash weight, resulting from dry lichen samples of standard size, can also be plotted against distance to obtain a rough estimate of the extent of a fallout zone. Occasionally, the accumulated dust and rock particulates do affect the lichens, for example around limestone quarries where the alkaline dust leads to a replacement of the normal lichen flora on nearby trees by lichen communities dominated by *Xanthoria, Caloplaca* and *Physcia*. Such genera are characteristic of eutrophicated (nutrient enriched) sites like farm roofs, the sea shore and bird perching stones (Gilbert, 1976).

7.7 Geobotanical prospecting

Certain flowering plants grow on soils rich in particular metals and are termed metallophytes. Recognition that such plants exist has led to the development of the geobotanical approach whereby a prospector, hoping to discover a mineral deposit, searches for particular plants as well as studying the geology (Shaw, 1990). Sometimes national herbaria are searched before fieldwork is undertaken. A supplementary approach is to determine the metal content of collected plant samples. Those with high levels may indicate the proximity of a mineral ore body. Crustose lichens growing on or near mineralised rocks may develop characteristically coloured thalli. For example, the rust-brown thalli of *Tremolecia atrata*, *Lecidea lapicida* forma *ochracea*, *Acarospora sinopica* and *Porpidia macrocarpa* variety *oxidata* are often indicators of iron mineralisation whereas green thallus margins or dark green fruit bodies are found, respectively, in *Lecanora cascadensis* and *L. polytropa* when these grow on copper-rich rocks.

Leafy and shrubby lichens such as *Peltigera* and *Cetraria* are often used in geobotanical prospecting because they are simpler to sample and like crustose lichens accumulate metals. Thus lichen samples collected near serpentine rocks are often found to accumulate high levels of nickel. Around Contwoyoto Lake, Northwest Territories, high levels of copper were found in *Cetraria* during a survey of lichen metal levels in northern Canada (Tomassini and others, 1976). The significance of this became apparent when a mining company independently discovered a copper ore deposit in the area. However, the use of lichens for geobotanical prospecting is limited unless considerable information is available on the normal background levels of each element in a particular lichen species (Puckett & Finegan, 1980). Furthermore, rural environments are becoming progressively contaminated by air pollution originating far away (transboundary). This can complicate interpretation of geobotanical data. Surveys of the metal content of *Hypogymnia physodes* in Denmark indicate that background atmospheric deposition is significantly higher in the eastern and southeastern parts of the country than in the western and northern parts. In other Scandinavian countries, such as Sweden, deposition is least in northern areas and increases in a southwesterly direction (Pilegaard and others, 1979).

8 Transplant studies

There is a zone in the centre of many urban and industrial areas where no lichens are found and this is often referred to as a lichen desert. In lichen deserts or in surrounding areas where the lichen flora is poor, transplants provide a way to study the effects of current air pollution levels on lichen survival or to assess what substances are being emitted at different times of year.

8.1 Effects of urban environments

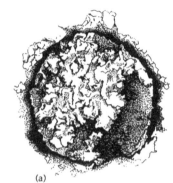

(a)

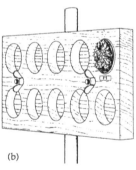

(b)

Fig. 20. A technique for transplanting lichens growing on tree bark. (a) a bark plug, 4 cm in diameter bearing a thallus of *Parmelia caperata* which has been cut from a tree and is ready for transplantation to a new tree. (b) A system used in Germany in which bark plugs are inserted in wooden exposure 'plates' when no suitable host trees can be found (from Brodo, 1961 and Schonbeck & van Hut, 1971).

Brodo was the first to devise a method for transplanting leafy lichens as part of an early but comprehensive study of the effects of city air pollution on a lichen flora (Brodo, 1961, 1968). Bark cores 4 cm in diameter were removed from oak trees in an unpolluted area of Long Island, New York. The bark plugs with their attached *Parmelia caperata* were affixed with grafting wax to host trees at varying distances from the city. After four months, thalli close to the city showed damage, turning first yellow and then white. At greater distances, this rather pollution-sensitive species was correspondingly less affected. A development of the technique was used in Westphalia, Germany. Bark cores of similar size to those in the Long Island study, and colonised by *Hypogymnia physodes*, were cut out using a drill and keyhole saw. They were inserted into recessed holes on a rectangular wooden block which was attached to a post 1.5 m above the ground (fig. 20) (Schonbeck & van Hut, 1971). Damage was evaluated photographically over a period of six weeks. This provided a picture of air quality in the overall area and around particular industrial emission sources.

More recently, *Hypogymnia physodes* has been used in studies around Budapest. In this case thalli were removed from trees in the rural control area and stitched to squares of felt (8 x 8 cm). Six to eight thalli were attached to each piece of felt and the transplants were affixed to trees at 50 points in the city. Thalli transplanted to the proximity of a sulphuric acid factory soon exhibited yellowish or reddish marginal lobes. Lichens transplanted close to main roads showed more serious discoloration than those attached to trees in quiet streets. When lichens were transplanted to opposite sides of trees, those on the side facing the road were more severely damaged (Farkas and others, 1985).

Hypogymnia physodes grows abundantly on the lower branches (often dying or dead because of shading from above) of plantation spruce trees. By taking such branches and attaching them to trees in urban industrial areas, it is possible to transplant large numbers of thalli. In Finland, branches covered with *Hypogymnia* were placed near pulp mills and fertiliser factories, and the impact of the gaseous air pollutants was assessed by looking at the ultrastructure of the lichen algae using a transmission electron microscope (Holopainen, 1984b). Thalli were collected after periods of from one to 40 weeks.

Those from the more polluted sites showed changes after only one week. In general, the speed with which symptoms developed, and their severity, related to distance from the source. Ultrastructural changes always preceded visible ones, and changes detectable by the light microscope (for example, the percentage of dead or plasmolysed cells) appeared slowly if at all in the less polluted areas. As different ultrastructural symptoms are associated with pollution by sulphur, fluorine and nitrogen compounds, electron microscopy is especially valuable for diagnosing the pollutant causing damage to such lichen transplants.

8.2 Monitoring airborne metals

An epidemiological survey in Armadale, Scotland revealed higher than normal numbers of lung cancer cases, especially in an area near the town's steel foundry. Leafy lichens were used to monitor metal contamination of the environment. The advantage of using transplants in this study included the facts that a high-density sampling grid could be set up, and that the cost of using lichens was very modest compared with the large number of mechanical air samplers that would otherwise have been required. Small branches covered with *Hypogymnia physodes* were tied with weather-resistant string onto plastic-covered wire which was then fixed to bamboo poles at a height of 2 m above the ground. The transplants were located in the back gardens of houses that faced the foundry. Transplants were left in place for two months, a period which preliminary studies showed to be optimum for metal accumulation by lichens (Gailey & Lloyd, 1986). Samples were collected regularly for 16 months and analysed for their metal content. Results revealed two main areas of metal deposition, one close to the foundry and a second smaller area in the north of the town (fig. 21). The latter was unexpected but wind tunnel experiments subsequently indicated that the local topography caused pollution from the foundry to be channelled into this area. This finding was of particular interest as a small 'cluster' of lung disease occurred in this part of the town (Gailey and others, 1985).

In another study using *Hypogymnia physodes*, transplants were used to monitor the effects of changing from open hearth to electric-arc furnaces at a steelworks in Frederiksvaerk, Denmark. Metal pollution in the general area was still severe after the change, although lichen transplants revealed a reduction in metal deposition at stations within 500 m of the emission sources (Vestergaard and others, 1986).

Transplantation studies need not be limited to leafy lichens. In Israel, twigs covered with the shrubby lichen *Ramalina duriaei* were transplanted to contaminated sites in a series of monitoring studies. One of these involved assessing the impact of the first coal-fired electricity generating station in Israel at Sharonim. Samples were transplanted to ten sites at varying distances from the power station and metal levels were determined both before start-up and periodically afterwards. The lichen data indicated that the regional concentration of

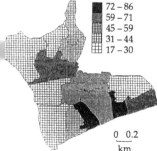

72 – 86
59 – 71
45 – 59
31 – 44
17 – 30

0   0.2
‾‾‾‾
km

Fig. 21. The zinc contents of *Hypogymnia physodes* (μg g^{-1} dry weight) transplanted to 47 locations in Armadale, a small industrial town in central Scotland (Gailey and others, 1985).

chromium had increased since operation began. At one site, high concentrations of nickel and chromium in transplanted lichens were clearly linked with the operation of the generating station and related to prevailing meteorology. The progressive decrease in zinc levels in the lichens over the period of operation of the station was explained in terms of a reduction in the use of Zn-containing sprays in nearby citrus and pecan orchards (Garty & Hagemeyer, 1988). Finally, the increase in lichen lead levels since start-up of the station reflected the greater use of automobiles in recent years (Garty, 1988). This last feature was investigated further by transplanting *Ramalina duriaei* to five sites exposed to varying levels of pollution from motor vehicle exhausts. The adenosine triphosphate (ATP) levels and chlorophyll concentrations were measured in the lichen samples. In all cases, the proportional decline in ATP concentration was greater than chlorophyll degradation. For example, samples transplanted to a site at Abou Kabir exhibited an 8% drop in chlorophyll *a* but a 50% fall in ATP as compared with controls. This suggests that ATP concentrations provide a sensitive measure of the effects of pollution on lichen transplants (Kardish and others, 1987).

8.3 Accumulation from contaminated water

Recent studies have shown that the efficient lichen uptake systems can be exploited to monitor metal contamination of rivers. Samples of *Parmelia praesorediosa*, which is abundant in southwestern Louisiana, were collected from a remote site near the Gulf of Mexico. After washing with distilled water and air drying, 5 grams of the lichen were packed into nylon mesh sacks and placed in PVC tubes (2.5 x 30 cm) with holes drilled in all sides, sealed with a cap at each end. The tubes were submerged for two weeks in a tributary of the Calcasieu River at ten sampling locations. The samples were retrieved, washed, dried and analysed. The pattern of Zn, Cu, Cr and Cd accumulation was very similar. Metal levels increased upstream to a maximum value at site 5, which was located just inside an industrial drainage ditch (fig. 22). Site 4 just downstream of this showed unexpectedly low values, probably owing to the pattern of current flow in the tributary. Upstream of site 5, metal levels dropped dramatically. Remarkably high levels of Hg were also found and this is thought to be related to contamination from bottom sediments. Hg has been used for over 40 years by the local chlor-alkali industry which discharges effluent at station 5 (Beck & Ramelow, 1990).

Lichen material has also been used sucessfully to construct lichen-modified carbon paste electrodes with potential as electrochemical biosensors (Connor and others, 1991). Studies are in progress in which lichen material is suspended in metal-contaminated waters. The retrieved lichen will then be incorporated into carbon paste electrodes. By determining quantitatively the metals that become adsorbed to the lichen, the researchers hope to evaluate directly the levels of biologically available soluble metal ions in rivers. These recent developments extend the value of lichens as biomonitors to a new medium.

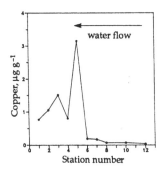

Fig. 22. Levels of copper in samples of *Parmelia praesorediosa* submerged for 2 weeks at various locations along the Bayou d'Inde, Louisiana (from Beck & Ramelow, 1990).

PLATE 1

Lichens of polluted areas

1. *Desmococcus viridis* (an alga)

2. *Lecanora conizaeoides*

3. *Lepraria incana*

4. *Buellia punctata*

5. *Diploicia canescens*

6. *Lecanora expallens*

7. *Xanthoria parietina*
   a: exposed form
   b: shade form

8. *Cladonia coniocraea*

9. *Cladonia macilenta*

10. *Lecanora dispersa*

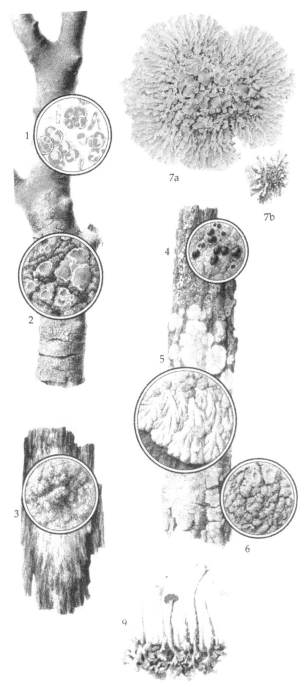

PLATE 2

Lichens of moderate pollution

1. *Hypogymnia physodes*

2. *Ramalina farinacea*

3. *Evernia prunastri*

4. a: *Physcia adscendens*
   b: *Physcia tenella*

5. *Lecanora chlarotera*

6. *Foraminella ambigua*

7. *Platismatia glauca*

8. *Lecidella elaeochroma*

9. a: *Parmelia sulcata*
   b: *Parmelia saxatilis*

10. *Parmelia glabratula*
    green (wet); brown (dry)

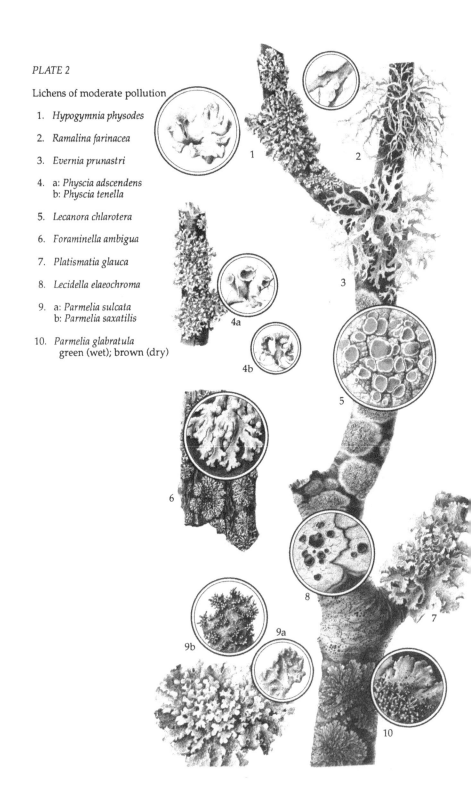

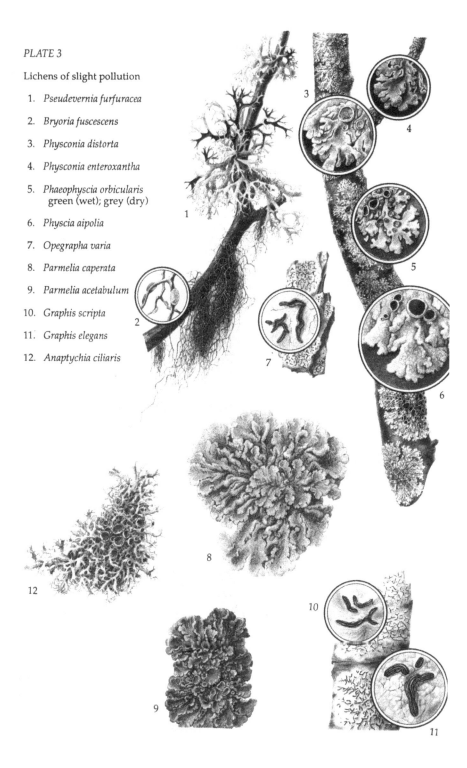

PLATE 3

Lichens of slight pollution

1. *Pseudevernia furfuracea*
2. *Bryoria fuscescens*
3. *Physconia distorta*
4. *Physconia enteroxantha*
5. *Phaeophyscia orbicularis*
   green (wet); grey (dry)
6. *Physcia aipolia*
7. *Opegrapha varia*
8. *Parmelia caperata*
9. *Parmelia acetabulum*
10. *Graphis scripta*
11. *Graphis elegans*
12. *Anaptychia ciliaris*

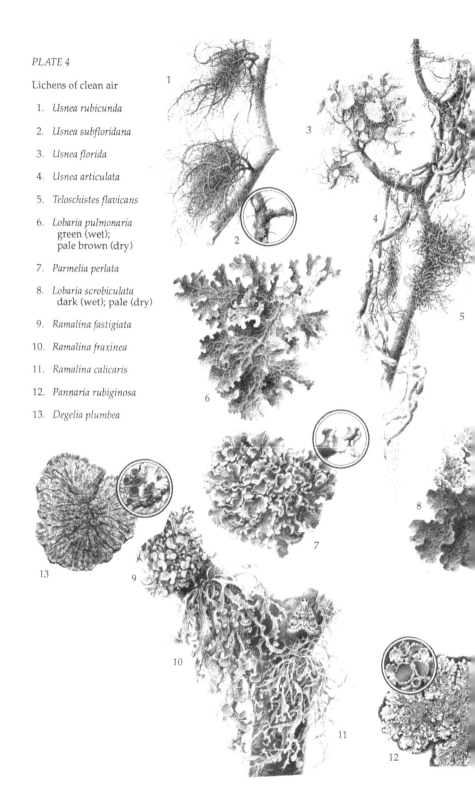

PLATE 4

Lichens of clean air

1. *Usnea rubicunda*

2. *Usnea subfloridana*

3. *Usnea florida*

4. *Usnea articulata*

5. *Teloschistes flavicans*

6. *Lobaria pulmonaria*
   green (wet);
   pale brown (dry)

7. *Parmelia perlata*

8. *Lobaria scrobiculata*
   dark (wet); pale (dry)

9. *Ramalina fastigiata*

10. *Ramalina fraxinea*

11. *Ramalina calicaris*

12. *Pannaria rubiginosa*

13. *Degelia plumbea*

9 Radioactive elements

Lichens have been used successfully to monitor contamination by radioactive elements derived from atmospheric nuclear bomb testing, the crashing of nuclear-powered satellites, and accidents such as that at the Chernobyl nuclear powered electricity generating station. Emissions from such sources may include radioactive isotopes, such as [137]caesium, of normally non-radioactive elements, or isotopes of radioactive elements like plutonium. The latter does not occur naturally, but forms during fission or fusion reactions as a mixture of radioactive isotopes such as [239]plutonium and [240]plutonium. Lichens accumulate these radioactive isotopes but the quantification of the transuranium elements in particular is difficult and time consuming. However, such research is of key importance because of the important information gained.

9.1 Atmospheric nuclear bomb testing

The efficiency with which lichens accumulate radioactive isotopes from the air has been known since the 1960s, following the period from 1950 to 1962 when atmospheric testing of nuclear bombs was widespread. It has been estimated that about 2.2×10^{17} Becquerels (Bq) of plutonium were released from the 400 or so tests. After each explosion, released plutonium remained in the stratosphere for about a year. A close relationship has been observed between the estimated fallout of radioactive isotopes of plutonium and its concentration in *Cladonia* mats (Holm, 1977; Holm & Persson, 1978) (fig. 23).

The importance of the lichen–reindeer–man food chain has also been realised since the 1960s (Pruit, 1963; Aberg & Hungate, 1967). Studies on ^{137}Cs, released during nuclear explosions, revealed that Lapps from Scandinavia and Inuit from northern Canada and Alaska had up to five times as much ^{137}Cs in their bodies as nearby people who did not depend on reindeer or caribou for food. Male Lapps had higher ^{137}Cs levels than female Lapps as they ate more reindeer meat. It is estimated that a reindeer consumes 2–5 kg of dry lichen per day during winter and browses some 2000 m^2 of *Cladonia* mat during this time (Holleman and others, 1979; Slack, 1988). The peak levels in Lapps are found in early summer (fig. 24).

The ^{137}Cs is accumulated by lichens in two ways. The first is associated with the direct transfer, to the lichen, of small particles to which the radioactive material is adsorbed (dry deposition), and the second involves the capturing of these particles by cloudwater. Soluble material dissolves in the water droplets in the clouds and in this way the caesium ions eventually fall onto the lichen in rain (wet deposition). Recent evidence suggests that in lichens the rate of uptake from dry deposition is often less than 10% of that from wet deposition (Smith & Ellis, 1990). Once taken up, ^{137}Cs acts in a

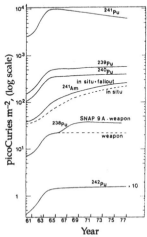

Fig. 23. The deposition of plutonium and americium at Lake Rogen, Sweden, estimated using samples of *Cladonia stellaris* collected annually. Accumulation up to 1965 was mainly from nuclear weapons testing. The increase in [238]Pu resulted from the crash of the US Snap 9A satellite. The [241]Am curve has two components, one from direct fallout and a second from [241]Am produced by the decay of [241]Pu (dashed line). The decline in the [241]Pu levels is due to this decay since [241]Pu has a half-life of only 14 years. The half-lives of the other isotopes are much longer (from Holm, 1977).

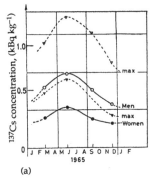

(a)

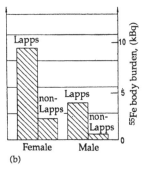

(b)

Fig. 24. (a) The seasonal
variation in [137]caesium
concentrations (average and
maximum values) during 1965
in Lapps (20–70 years old)
living near Funasdalen,
Finland, and (b) the [55]iron
levels in Lapps and non-Lapps
from the same area during
December 1965 (from Liden &
Gustafsson, 1967 & Persson,
1967).

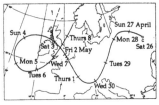

Fig. 25. The estimated path,
between 26 April and 8 May
1986, of the Chernobyl
radioactive cloud which
eventually crossed Britain,
where it divided. The main
body of the cloud then
proceeded northwards (from
Smith & Clark, 1989).

very similar way to potassium, being largely intracellular. The
concentration decreases with depth from the actively growing
tips of a *Cladonia* mat. The fact that part of the ^{137}Cs is particle-
bound may account for its longer retention time in lichens
than potassium which has similar chemical and physiological
properties. Thus, the biological half-life for ^{137}Cs has been
estimated to be about 17 years while that for potassium is
only around four years (Looney and others, 1986).

Female Lapps had higher levels of [55]Fe than males, in
contrast to the situation with [137]Cs. [55]Fe is another isotope
released by atmospheric nuclear testing. This high level
occurred even though the women eat less reindeer meat.
The higher uptake efficiency of the female intestine for iron,
which has to be replaced as a result of menstruation,
accounts for the different uptake pattern (fig. 24) (Liden &
Gustafsson, 1967; Persson, 1967).

9.2 Nuclear-powered satellites

Many satellites have been sent into orbit with
miniature nuclear generating plants to run the onboard
instruments. About 1.5×10^{16} Bq of plutonium have been
used in US satellites which will eventually re-enter the
atmosphere and burn up. The resulting fallout can be
monitored by analysing mats of *Cladonia* for any increased
plutonium content. This was done in the case of the
American Snap-9A satellite which released some 6.3×10^{14}
Bq of plutonium (fig. 23). However, attempts in northern
Canada to investigate the crash of the Russian Cosmos-954
satellite in January 1978 proved less successful. This was
because of significant background radioactivity derived
from nuclear explosions, including the Chinese test in
March 1978 which took place before the Canadian samples
could be collected and analysed (Taylor and others, 1979).

9.3 The Chernobyl accident

In April 1986, a large explosion occurred at the
Chernobyl nuclear-powered electricity generating plant in
the Ukraine. This was not the world's first nuclear reactor
accident. There was a serious fire at the Windscale (now
Sellafield), UK nuclear facility in 1957, but the scale of the
Chernobyl accident was much larger and monitored in
greater detail. The explosion and subsequent fires released a
plume of radioactive particles for many days before effective
controls were implemented. The paths followed by the
emitted radioactive materials were complex (fig. 25) (Smith
& Clarke, 1986, 1989). Furthermore, the fallout showed
extreme geographic variability depending on the pattern of
rainfall: heavy precipitation led to local hot spots of
contamination (Hohenemser and others, 1986). A significant
amount of material was eventually deposited over
Scandinavia, with serious economic consequences for the
Lapp reindeer herders. Unacceptably high levels of [137]Cs
were, and still are, found in the reindeer meat of many

animals. Levels in excess of 10,000 Bq of ^{137}Cs per kilogram were measured in some reindeer, with a legal limit for sale of only 300 Bq kg[-1] (Mackenzie, 1986; O'Clery, 1986). Such animals cannot be sold but fortunately those which are severely contaminated come from fairly well delimited areas so that most Lapps can still use their deer for food. However, consumer concern about radioactivity in reindeer and other deer meat has led to a fall in demand. In the next few years, this may affect the economy of Lapps and upland deer farmers in other areas of Europe.

The impact of the Chernobyl accident has also been monitored in Poland, where there was already an ongoing study of radioisotopes in lichens including [226]radium [derived from uranium] emitted from the burning of lignite and other fossil fuels. The lichen *Umbilicaria* was used in this programme and the greater [226]Ra content in samples collected above 800 m reflected the increase in rainfall at higher altitudes (Kwapulinski and others, 1985). As a result of this study, ^{137}Cs data were available to compare with those from lichens collected after the Chernobyl accident. Analyses revealed that post-Chernobyl samples contained ^{137}Cs levels within the range 471–36,630 Bq kg[-1]. This was a startling increase (up to 165-fold) over material monitored seven years earlier when a range of only 18–226 Bq kg[-1] had been detected. Exceptionally heavy rainfall in the area, at the time the pollution cloud passed over, again accounts for the high contamination level found in the lichens (Seaward and others, 1988 and personal communication). Similar results have been recorded for Austria and somewhat lower levels (1,070–14,560 Bq kg[-1]) for a range of lichens collected in Greece (Turk, 1988; Papastefanou and others, 1989). In Britain, particularly in Wales and Cumbria, there are continuing restrictions on the movement and sale of upland sheep which have unacceptably high levels of ^{137}Cs derived from the accident. The fungi associated with the roots of heather (mycorrhizas) and related plants accumulate, recycle and transfer the ^{137}Cs to the heather grazed by the sheep.

The Chernobyl pollution cloud arrived in Canada 11 days after the accident and its movement across this country provided a unique opportunity to validate a model describing the uptake of radioactive isotopes by lichens. Samples of *Cladonia rangiferina* were collected on an east–west transect across the maritime provinces. A deposition velocity of 1.1 cm sec[-1] gave good agreement between the model and experimental results. Using this result, combined with information on atmospheric residence times, it was possible to calculate that the radioactive cloud passed over Canada at a height of 10,000 m (Smith & Ellis, 1990). Here again, lichens proved to be excellent collectors of atmospheric aerosols owing to their high surface-to-mass ratio, slow growth rates and the fact that they derive their nutrients from atmospheric sources. These scavenging properties make lichens very cost-effective monitoring devices for radioactive isotopes (Smith & Ellis, 1990).

10 Invertebrate fauna

The surfaces of lichens which grow on rocks and trees provide resting sites for several species of moths which exhibit crypsis to these plants. In other words, when on a lichen they are camouflaged and much less visible to predators. Furthermore, a rich fauna of small invertebrates depend on lichens for food or shelter. In relation to pollution monitoring, two aspects are of particular interest: the first concerns melanism in moths and the second microarthropods and other small organisms (including protozoa).

10.1 Melanism in moths

Fig. 26. The *carbonaria* and *typica* forms of the peppered moth *Biston betularia* resting on a lichen-covered trunk and showing the better crypsis of the *typica* form.

Among senior schoolchildren and other students, the peppered moth *Biston betularia* is the best known of the larger moths as its study is often part of the biology curriculum. Two forms of the moth occur: the pale *typica* and the dark *carbonaria*. These are cited in discussions on natural selection, the central process of Darwin's theory of evolution. The pale form has white wings liberally speckled with black. In 1848, a predominantly black, *carbonaria*, form was recorded in Manchester (fig. 26). By the end of the century 98% of the peppered moths collected in this area were of the dark (melanic) form. This led to the concept of industrial melanism and the proposition that the *carbonaria* form survived bird predation better on the trunks of bare smoke-blackened trees in industrial areas, whereas the pale *typica* form was more advantaged in rural areas where trees were lichen-covered. Experiments, in which both forms were pinned to tree trunks and bird predation was observed, seemed to confirm the hypothesis (Kettlewell, 1959, 1973).

Recent studies on this moth reveal that the phenomenon of melanism is more complex and far more interesting than orginally thought (Majerus, 1989). Thus, in a sample of collected moths, most will be either pale or dark (having either the dominant dark allele or the pale recessive allele of the gene controlling colour). However, the *carbonaria* forms vary widely in coloration, and the colour of the intermediate *insularia* forms is affected by at least three additional alleles. Furthermore, the earlier bird-predation studies were based on the assumption that the moths rested on tree trunks but recent research indicates that moths prefer to rest on twigs adopting specialised attitudes. On twigs, they normally rest against leafy lichens, copulate for up to 24 hours and then deposit their eggs below lichens or in cracks if these epiphytes are absent (Majerus, 1989).

In Cardiff, an urban area, mating pairs involving *typica* moths were highly contrasted against the uniformly dark branchlets but on the trunks pairs involving *insularia* or *typica* forms were more cryptic than those involving

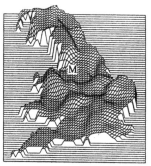

Fig. 27. Differences in *carbonaria* frequencies in the peppered moth, *Biston betularia*, between the 1970s and 1980s. The vertical scale represents the decline in melanic frequency (or increase in *typica* frequency). The 'ridge' between Manchester (M) and the west coast of North Wales indicates that melanic frequencies declined particularly sharply here over the period (Cook and others, 1990).

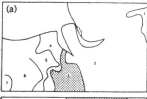

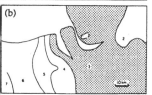

Fig. 28. The lichen zones recorded in 1973 (a) and 1986 (b) on the transect from Manchester to North Wales. Along this transect a marked change in the proportion of melanic to typical forms of the peppered moth was observed between the two dates. Note the disappearance by 1986 of zone 1, 'a lichen desert' in the east and the extension of zone 3 (shaded). This zone is characterised by abundant growth of the pollution-tolerant *Lecanora conizaeoides* which spreads up the trees from the trunk base to which it is confined in zone 2 (Cook and others, 1990).

carbonaria forms. In rural Somerset, melanic forms were generally less cryptic, but there were exceptions: for example, when a melanic was the lowermost moth of a mating pair on a twig. There was little predation by birds at night, but during the day high mortality of the *carbonaria* form was observed at both sites. Predation of all forms was higher at the Cardiff site, with *carbonaria* x *carbonaria* mated pairs having a higher mortality than expected in relation to its cryptic coloration. In Somerset, *typica* x *typica* pairs had a relatively low mortality from predation which seemed to confirm the effectiveness of its camouflage (Liebert & Brakefield, 1987). Clearly, more studies are required to confirm Kettlewell's hypothesis of differential survival.

The effect of the implementation of the 1956 and 1968 Clean Air Acts in Britain has been revealed by studies on the proportion of the dark *carbonaria* form relative to pale *typica* form of the peppered moth that occur in different parts of Britain. This has dropped, in Cambridge, from over 90% to less than 40% (Majerus, 1989). Along a transect from urban Manchester to North Wales, an eastward shift in the cline in frequency of the dark form has been explained in terms of falling pollution levels leading to the spread of the pollution-tolerant lichen, *Lecanora conizaeoides*, up the trees to cover both trunks and branches at the eastern end of the transect (fig. 27). There has also been an increase in lichen diversity in the central portion of this transect (fig. 28) (Cook and others, 1990). There are many other night flying moths that have melanic forms in Britain and also in the industrial areas of other European countries, Canada and the USA. Some, like the Scalloped Hazel moth and the Pale Brindled Beauty moth, have been the object of research (Bishop & Cook, 1975), but others deserve closer study (see next section). It would be interesting to study the extent of the change from melanic to typical forms in such moths as air pollution levels continue to fall.

10.2 Microfauna

Lichens are often used as biomonitors of air pollution, but the value of their associated microfauna is less widely recognised. The numbers of bark-living invertebrates around Newcastle upon Tyne declined in parallel with the disappearance of lichens and bryophytes on trees as a result of increasing smoke and sulphur dioxide (Gilbert, 1971). Studies on mite communities on trees in areas affected by air pollution in Belgium suggested that the oribatid mite *Humerbates rostrolamellatus* is a fast and reliable bioindicator of air quality (Andre and others, 1982). The association of oribatid mites with lichens is well established and the changes that occur in relation to air pollution probably reflect reduced food and microhabitat resources (Seyd & Seaward, 1984).

In a recent study of the microfauna of lichens around Belfast, Ireland, lichen samples were collected along a series of transects from the city centre. Small pieces were incubated in 5 ml of sterile water and examined

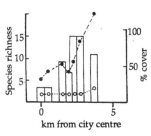

Fig. 29. The lichen community on trees in relation to the diversity of the associated microfauna along a transect from the city of Belfast, Ireland to Glencairn. Blocks: % cover by the pollution-tolerant *Lecanora conizaeoides*. (O) is the increase in lichen diversity and (●) the microfaunal species richness. The microfaunal richness increases sharply as foliose lichens begin to colonise the trees along with *Lecanora conizaeoides* (from Roberts & Zimmer, 1990).

microscopically after one hour and after 4 and 9 days. The diversity of both the lichens and the associated microfauna increased with distance from the city centre. Except in the very centre of the city, mites, springtails, rotifers and nematodes were widely distributed and therefore not useful bioindicators. In contrast, tardigrades such as *Macrobiotus hufelandii* and the cosmopolitan carnivorous *Milnesium tardigradum* were absent from the more polluted areas and had potential as bioindicators as they were considered sensitive to air pollution. Similarly certain ciliates, and flagellates were widespread, occurring in lichens from the most polluted areas. Others, such as the ciliates *Glaucoma scintillans* and *Woodruffia lichenicola*, and the rhizopods *Euglypha* species and *Trinema lineare*, were restricted to unpolluted sites. The observed increase in microfaunal diversity with distance from the city centre was largely accounted for by the enhanced number of protozoan species associated with the lichens (fig. 29). These under-studied animal associates of lichens may thus be valuable biomonitors of air quality (Roberts & Zimmer, 1990).

11 Identification

11.1 Identification key

K.1

K.2

(a)

(b)

(c)

K.3

K.4

This key covers the lichens illustrated in plates 1–4 plus a few species that have a similar appearance. They all grow on trees and have been selected because they span the range of tolerance to air pollution, particularly by sulphur dioxide. Other species may be found, so that it is important to ensure that both the description and illustration match the lichen being examined. If there is any uncertainty or if identification of other species is desired, then Dobson (1992) should be consulted.

This key has been designed to be used with only a hand lens or good close-up vision. However, for some lichens, chemical tests and examination of the spores within the fruit bodies are necessary. The chemical tests are coded in the key (for example K+y, C+o). The codes are explained, and the tests are described, at the end of the key (p. 52). Leafy or shrubby lichens can generally be identified without examining the spores. Scientific terms used are explained in more detail on pp. 54–60.

In very polluted city centres there may be no lichens growing on the trees so that only brown or soot-blackened bark is evident. In less polluted suburbs or on shaded parts of tree trunks in rural areas crusts of green algae, especially *Desmococcus* (pl. 1.1), may be found. Such algae form bright pea-green or dark spinach-green crusts. Lichen crusts are never this colour. The thallus of *Lecanora conizaeoides* (pl. 1.2) might initially be mistaken for an alga, but it is grey-green and forms a 2–3 mm-thick crust which is usually cracked like miniature crazy paving. Furthermore, this lichen usually has round, pale yellowish green fruit bodies on the surface which can be seen with the help of a hand lens or by viewing closely with the naked eye.

1   Lichen thallus so closely attached to the tree that it cannot be removed without taking the bark as well. This type of lichen forms a crust on the bark [crustose] (K.1) or consists of a powdery mass attached to the bark [leprose] (K.2).   crustose and leprose lichens 2

–   Lichen thallus not closely attached to the tree bark. Such lichens can be picked off by hand in the case of shrubby [fruticose] lichens (K.3) or collected using a knife to separate the lower surface of the lichen from the bark in the case of leafy [foliose] lichens(K.4).
        fruticose and foliose lichens 4

2   Thallus a powdery mass of whitish grey, green-grey, bright green or bright yellow powdery soredia (K.1, pl. 1.3)   3

–   Thallus a crust inseparably attached to the bark   39

3   Thallus a loose powdery whitish grey or green-grey
     mass growing on shaded areas of bark or near the base
     of tree trunks (pl. 1.3)   *Lepraria incana*

     (A similar but apple-green leprose lichen, *Lepraria lobificans*, is
     common in western Britain while a conspicuous bright yellow
     leprose lichen, *Lepraria candelaris*, grows in the cracks of the
     bark on oak and other trees in unpolluted areas. Neither is
     illustrated on the chart.)

–   Thallus a flat yellowish green crust covered with
     powdery soredia (K–, C+o) (pl. 1.6)   *Lecanora expallens*

4   Thallus fruticose (shrubby) (K.3) with round or strap-
     shaped branches that are erect or pendant and attached to
     the bark by their bases or via a basal cluster of lichen
     scales   5

–   Thallus foliose (leafy) (K.4) with flat leaflike lobes more
     or less pressed against the bark surface   19

5   Branches tubular and hollow, upright, usually less than
     3 cm tall and often with a cluster of scales at the base.
     The upright structures (podetia, K.3a) may have red or
     brown fruit bodies at the tip. Common on tree bases
                                                genus *Cladonia* 6

–   Branches round or strap-shaped but solid, upright or
     pendant and attached directly to the bark by the base
     (K.3b, c)   7

6   Podetia with red fruit bodies at tip, and at the base
     clusters of small scales with rather ragged margins.
     Grey-green (K+y, P+o) (pl. 1.9)   *Cladonia macilenta*

–   Podetia with brown fruit bodies at the tip, often horn-
     shaped (not as pointed as previous species). This lichen
     is dull yellowish green to green (K–, P+r) (pl. 1.8)
                                              Cladonia coniocraea

     (Other species of *Cladonia* may be encountered, including
     Cladonia pyxidata (the pyxie cup lichen) with cup-shaped
     podetia (K–, P+r) (K.3a), and *Cladonia squamosa* which is like
     Cladonia macilenta but has scales all the way up the podetia
     and brown fruit bodies (K–, P–). These are not illustrated.)

7   Thallus consisting of round, solid and often hairlike
     branches   8

–   Thallus consisting of distinctly flattened branches   13

8   Branches bright orange-yellow, less than 5 cm tall. Very
     rare in very unpolluted sites in southern Britain (pl. 4.5)
                                              Teloschistes flavicans

–   Branches green, grey-green, reddish green or brown   9

K.5

9   Thallus consisting of hairlike grey-brown to blackish
     branches forming a tangled mass (K.5) (pl. 3.3)
                                              Bryoria fuscescens

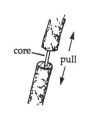

K.6

– Thallus consisting of green, grey-green or reddish green branches with a strong white or pinkish core visible when main branches are stretched (K.6) (the beard lichens)   genus *Usnea* 10

10 Thallus with conspicuous main branches that are often over 10 cm long with constrictions along them so that they appear like a string of sausages. The lichen forms a rather loose mass with relatively few side branches and often no obvious base. Rare but locally abundant in very unpolluted sites in western and southern Britain (medulla K–, P+r) (pl. 4.4)   *Usnea articulata*
– Thallus without constrictions, consisting of a tuft of branches coming from a distinct base   11

11 Branches dull red to reddish green. Locally common in southern and western regions in unpolluted sites, rarer northwards (medulla K+y, P+o) (pl. 4.1) *Usnea rubicunda*
– Branches green to grey-green   12

12 Fruit bodies abundant, large, conspicuous and disc-like on ends of branches. Common in old woodlands in unpolluted sites in southern and western Britain (medulla K+y, P+o) (pl. 4.3)   *Usnea florida*
– Fruit bodies few or usually absent; main stem often with blackened area for a short distance above the point of attachment to the tree. This lichen is spreading in Britain since sulphur dioxide levels have fallen and often very small thalli are found on ash and other deciduous trees (medulla K+y, P+o) (pl. 4.2)   *Usnea subfloridana*

(Several other green species of *Usnea* occur but they are difficult to distinguish from this species and are probably of similar pollution sensitivity.)

13 Strap-shaped branches green or grey on upper surface but whitish or black on lower surface so that the two sides of the lichen look different   14
– Strap-shaped branches green on upper and lower surfaces (though the latter may be slightly paler green). Overall, both sides look similar   genus *Ramalina* 16

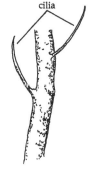

K.7

14 Conspicuous long hairs (cilia) (K.7) projecting from the margin. These cilia often entangle the upward growing branches. These branches may be 5 cm or more long (fruit bodies with black centres are common in unpolluted areas) (pl. 3.12)   *Anaptychia ciliaris*

(A smaller lichen, *Physcia leptalea*, not illustrated, may also be found. This too has marginal cilia but they are shorter. The branches are more flattened and strap-shaped and usually not more than 2–3 cm long. The fruit bodies are infrequent and have grey or black centres.)

– No cilia on the margins   15

15 Upper surface grey, lower surface black (often white in
    smaller thalli or on younger parts of thalli). Common on
    trees mainly in upland regions (K+y) (pl. 3.1)
                                          Pseudevernia furfuracea
–   Upper surface pale green, lower surface whitish (pl. 2.3)
                                          Evernia prunastri
    (If you arrive at this point in the key and are looking at a
    lichen with brownish green upper surface and white lower
    surface which becomes pale brown near the margin of the
    thallus, see *Platismatia glauca*, couplet 21, and the
    accompanying note.)

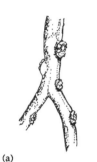

16 Fruit bodies conspicuous and very numerous forming at
    the ends of the short and bushy branches (pl. 4.9)
                                          Ramalina fastigiata
–   Fruit bodies absent or when present not numerous and
    tending to be at the sides rather than ends of branches 17

(a)

17 Margins of the strap-shaped branches with small
    yellow-green areas (K.8a) where powdery soredia are
    formed (pl. 2.2)   *Ramalina farinacea*
    (Another species forming soredia, *Ramalina canariensis*, is
    found on trees. It has short wrinkled and puffed-up lobes
    which are split at the tips where the soredia occur (K.8b). It is
    not illustrated in the plates.)
–   Margins of the strap-shaped branches without soredia;
    branches wide and wrinkled **or** narrow and channelled 18

(b)
K.8

18 Branches wide and irregular with wrinkled surface; fruit
    bodies scattered along the margin (pl. 4.10)
                                          Ramalina fraxinea
–   Branches channelled; fawn coloured fruit bodies on or
    near tips but these are not numerous and do not
    dominate the lichen (as they do in *Ramalina fastigiata*)
    (pl. 4.12)   *Ramalina calicaris*

19 Thallus green, grey or brown, consisting of narrow or
    broad lobes which may be pressed against the bark, or
    broad and more upright but the overall effect is still
    leaflike (foliose)   20
–   Thallus bright yellow to orange, foliose (K+p) (pl. 1.7)
                                          Xanthoria parietina
    (In shade the lobes may be grey-yellow but the central disc of
    the fruit bodies is still yellow and gives the K+p reaction. There
    are also two smaller species, *Xanthoria polycarpa* (in which the
    fruit bodies are usually so numerous that the underlying
    thallus, about 2 cm across, is difficult to see) and *Xanthoria
    candelaria* (in which the rather upright growing lobes of the
    thallus, about 3 cm across, have yellow soredia at the tips, and
    fruit bodies are usually absent). Neither is illustrated. All the
    species of *Xanthoria* are very common in habitats receiving
    nutrient enriched dust, such as roadsides near arable fields,
    farms, cement factories, and sewage works.)

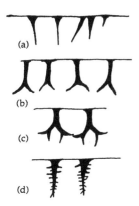

(a)

(b)

(c)

(d)

K.9 Types of rhizinae

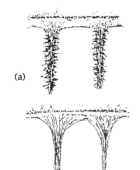

(a)

(b)

K.10 Typical rhizinae of *Peltigera*

20 Individual lobes of the thallus large, about 1 cm wide, with upturned margins; lower surface either felty or smooth   21

–   Individual lobes medium or small, less than 1 cm wide, often attached to the substrate by black fungal strands called rhizinae (K.9)   23

21 Undersurface smooth, usually white in the centre becoming brown towards the margin; upper surface brownish green to grey, with erect crisp margins (pl. 2.7)
Platismatia glauca

(This lichen is often regarded as fruticose because only the centre of the plant is attached to the tree bark. However, the overall impression, particularly to beginners, is of a leafy lichen with crisp upward-growing edges. For this reason, it is keyed out here.)

–   Undersurface felty, thallus of broad lobes forming colonies that are often 10 cm or more across
genus *Lobaria* 22

(Lichens belonging to *Sticta* and *Pseudocyphellaria* key out here. These rare lichens have small white breathing pores (holes the size of a pinhead or larger) on the undersurface of the lichen and a brown or grey upper surface. There are also some species of *Peltigera* which normally grow on soil but sometimes grow on trees and will key out here. *Peltigera* has broad lobes, long white attaching strands (rhizinae) from the undersurface (K.10) and a blue-grey upper surface.)

22 Lobes bright green when wet and khaki green when dry; there are depressions on the upper surface which is wrinkled and ridged in between where soredia may develop. (Very sensitive to pollution and stress. This is one of the few lichens with a common name, the tree lungwort, because of the resemblance of the upper surface to the surface of a lung) (pl. 4.6)   *Lobaria pulmonaria*

–   Lobes yellow-grey to blue-grey; irregular, dark roundish areas stick up from near the lobe tips and bear coarse soredia or isidia. Found only in the north and west of Britain and apparently sensitive to acid rain (pl. 4.8)
Lobaria scrobiculata

23 Individual lobes of the thallus broad (between 3 mm and about 1 cm wide), forming colonies that may be large and circular in outline, but may form sheets or irregular colonies on the tree bark
genera *Parmelia*, *Hypogymnia*, *Degelia* and *Pannaria* 24

–   Individual lobes of thallus narrow (less than 3 mm wide), usually forming small circular lichens usually less than 5 cm in diameter
genera *Physcia*, *Phaeophyscia* and *Physconia* 34

(If you have any doubt, a look at the illustrations will help you decide which route to try first from this couplet.)

24 Thallus blue-grey, containing cyanobacteria. Hence a blue-grey layer beneath the surface of the thallus is seen if the lichen is torn or cut with a knife and observed with a hand lens. The lichens in this category occur only in unpolluted situations in west and north Britain   25

–   Thallus yellow-green, brown or pale grey (never blue-grey), containing green algae. Hence a bright green layer is seen beneath the pigmented upper surface if the lichen is torn or cut through with a knife. These lichens are often attached to the bark by rhizinae (K.9), especially in the central area of the thallus   26

dark fibrous margin

K.11

hollow centre

K.12

25 Thallus felty, forming large, grey circular patches (like rosettes) with a dark margin (K.11), up to 10 cm or more across. Fruit bodies dark rust-coloured with no grey margin around each. On old trees in unpolluted areas in the north and west of Britain and in Ireland (pl. 4.13)   *Degelia plumbea*

–   Thallus consisting of overlapping blue-grey scales, forming small rosettes. Fruit bodies tan-coloured, each having a grey margin. Of similar distribution to the last species (pl. 4.12)   *Pannaria rubiginosa*

26 Thallus with hollow lobes, each lobe looking inflated and having an air space between the upper grey part and the brown lower surface (K.12). This can be seen by tearing or cutting across a lobe and observing with a lens or closely with the naked eye (K+y, medulla KC+r, P+r) (pl. 2.1)   *Hypogymnia physodes*

(a)

–   Thallus with solid lobes   27

27 Thallus grey   28
–   Thallus brown or yellow-green   31

(b)

28 Thallus with powdery soredia   29
–   Thallus with peg-like or bun-shaped structures (termed isidia, K.13) projecting from the upper surface   30

(c)

K.13 Types of isidia

29 Soredia form on the edges of the lobes. The margins of lobes tend to be upturned. There are sometimes a few hair like cilia at the edge of the thallus (K.14). The black rhizinae are absent from the thallus margins, which are tan-coloured (K+y, medulla K+y, P+o) (pl. 4.7)   *Parmelia perlata*

–   Soredia form on white or very pale brown cracks which often form a net-like pattern that may cover the surface of the lobes. Underside dark brown and attached by black rhizinae almost to the margin (medulla and soredia K+y-r, P+r, C–) (pl. 2.9a)   *Parmelia sulcata*

cilia

K.14

(A few other species of grey *Parmelia* may be found in old woodlands. One has soredia in pinhead-like spots on the upper surface of the thallus. It is *Parmelia subrudecta* and has a C+r medulla.) (If you get to this point and your grey lichen fits neither the description nor the corresponding illustration, go back to couplet 23 and try the first route.)

30 Isidia peg-like or coral-like, grey with darker tips, growing on a thallus which is grey but has fine white cracks, especially near the margins. Although grey on sunlit branches, thalli may be green-grey in the shade. Undersurface dark brown, paler at margins (K+y, P+y, medulla K+y–r, P+o) (pl. 2.9b)   *Parmelia saxatilis*

–   Isidia bun-shaped (K.15), dark brown growing on a smooth bluish grey thallus (K+y, medulla C+r), not illustrated   *Parmelia pastillifera*

.15

(Another, similar species, *Parmelia tiliacea*, not illustrated, has brown-tipped isidia (K.16). It occurs in the east of Britain (K+y, medulla C+r).) (If you get to this point and your grey lichen fits neither the description nor the corresponding illustration, go back to couplet 23 and try the first route.)

16

31 Thallus yellow-green to apple-green   32
–   Thallus green-brown to dark brown   33

32 Lobes broad, up to 1 cm across, often wrinkled in the centre and forming large thalli 10 cm or more across; yellow-green soredia on much of the upper surface (P+o, medulla K+y) (pl. 3.8)   *Parmelia caperata*
–   Lobes narrow, about 2 mm wide, forming small circular patches less than 3 cm across with yellowish soredia confined to the centre of the thallus (pl. 2.6)
                                        Foraminella ambigua

33 Thallus with isidia projecting from upper surface; shiny, mid to dark brown, forming colonies closely pressed to the bark. Undersurface black in the centre but light brown at the margins where there are no rhizinae (medulla C+r, KC+r) (pl. 2.10)   *Parmelia glabratula*

(A common related species, *Parmelia subaurifera*, is often matt-brown and occurs on the younger branches. The isidia which are irregular in shape break off to leave scars which are very pale yellowish. It has the same chemical reactions.)

–   Thallus without isidia, dull lead-brown becoming more green-brown when wet. Undersurface light brown. Lobes wrinkled in the centre and rounded at margin. Occurs in eastern and southeastern Britain, becoming rarer northwards (medulla K+r, P+o) (pl. 3.9)
                                        Parmelia acetabulum

34 Thallus yellow-grey, forming circular patches up to about 3 cm across with yellowish soredia in the centre. Undersurface black, attached by rhizinae (this lichen resembles a small narrow-lobed *Parmelia*; see couplet 32) (pl. 2.6)   *Foraminella ambigua*

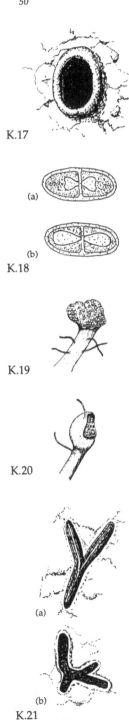

K.17

(a)

(b)

K.18

K.19

K.20

(a)

(b)

K.21

–   Thallus grey (sometimes greenish when wet), fruit
        bodies when present usually with grey or black disc
        surrounded by a grey margin (K.17) (spores within fruit
        body brown with one cross cell wall)
                    genera *Physcia* (K.18a), *Phaeophyscia* (K.18a) and
                                                    Physconia (K.18b) 35

        (Members of these genera are particularly common on bark
        receiving nutrient-rich dust from farms, cement works and
        sewage plants.)

35   Thallus without soredia or isidia; fruit bodies usually
        present   36
–   Thallus with soredia or isidia; fruit bodies usually
        absent   37

36   Thallus pale grey with a mottling of white spots on the
        upper surface, best seen with a hand lens. The fruit
        bodies have a black disc which is sometimes grey,
        especially when young (K+y) (pl. 3.6)   *Physcia aipolia*
–   Thallus grey to grey-brown (greenish when wet) with
        no mottling. The lobe tips look whitish when dry and
        the fruit bodies have a a grey disc (K–) (pl. 3.3)
                                                            Physconia distorta

37   Thallus with white medulla (revealed by tearing or
        cutting the thallus) and with soredia in small round
        patches on lobe surface or on lobe tips   38
–   Thallus with yellow medulla (pl. 3.4)
                                                    Physconia enteroxantha

38   Thallus with pale grey lobes not closely pressed to bark.
        Soredia borne on the lobes, behind the lobe tip (K.19)
        (pl. 2.4b)   *Physcia tenella*

        (In a similar species, *Physcia adscendens* (pl. 2.4a), the ends of
        the lobes are puffed up. These split and soredia form within
        the swollen helmet-shaped tips (K.20). Both have a K+y
        reaction.)
–   Thallus flat, with brownish grey (green when wet) lobes
        closely pressed to bark; soredia occur on roundish
        patches on the lichen surface (medulla K–) (pl. 3.5)
                                                    Phaeophyscia orbicularis

        (A similar species, *Physcia caesia* (pl. 2.4a) , has a medulla
        K+y.) (If you get to this point and have a brown or grey lichen
        which does not fit the descriptions or accompanying
        illustration for any of the above five species, go back to
        couplet 23 and try the second route, as you may have a
        species of *Parmelia*.)

39   Thallus with elongate grey or black fruit bodies that
        may resemble Chinese writing (K.21)   40
–   Thallus with round fruit bodies or lacking fruit bodies 42

40 Fruit bodies long, but the disc of each widening towards the centre so that it is rather boat-shaped (spores with 4–6 cross walls) (pl. 3.7) *Opegrapha varia*

(There are several other rather similar species of *Opegrapha* that occur on trees. The spores should be examined, and Dobson (1992) consulted, to make a certain identification.)

– Fruit bodies to 1 cm long, not widening at the centre, and resembling Chinese writing; spores elongate with many cross walls (often more than 10)   genus *Graphis* 41

K.22

41 Fruit bodies with margins having several parallel ridges when viewed with a lens (K.22) (pl. 3.11) *Graphis elegans*

– Fruit bodies with narrow unridged margins (K.23) (pl. 3.10) *Graphis scripta*

(Another lichen, *Opegrapha atra*, not illustrated, has elongate small (to 2 mm long) black fruit bodies which instead of being scattered and like Chinese writing are in closely spaced parallel lines. Its spores have 3 cross walls)

K.23

42 Thallus with round fruit bodies 43
– Thallus without fruit bodies 46

43 Fruit bodies with central disc and a surrounding margin the same colour as the thallus (looking like a miniature jam tart) (K.24), spores colourless without a cross wall genus *Lecanora* 44

– Fruit bodies with a disc but with a surrounding margin of the same colour which contrasts with the thallus (like a jam tart in which there is just the jam and no surrounding pastry) (K.25) 49

K.24

44 Thallus green to grey-green 45
– Thallus white to grey, usually with numerous fruit bodies having buff to red-brown discs (pl. 2.5) *Lecanora chlarotera*

K.25

45 Thallus with soredia, grey-green, rather thick (2–3 mm), and cracked, and fruit bodies with greenish buff to flesh coloured disc (P+r) (pl. 1.2) *Lecanora conizaeoides*

– Thallus lacking soredia, thin with scattered small fruit bodies having greenish grey to brownish discs (P–) (pl. 1.10) *Lecanora dispersa*

46 Thallus green or grey-green or yellow-grey 47
– Thallus grey to blue-grey 48

47 Thallus thick (up to 2–3 mm), grey-green and cracked (P+r) (pl. 1.2) *Lecanora conizaeoides*

– Thallus thin, green to yellow-green, covered with soredia (C+o) (pl. 1.6) *Lecanora expallens*

48 Margin of thallus smooth, distinct and fluted; soredia in central part only (K+y) (pl. 1.1)   *Diploicia canescens*
–   Margin of thallus powdery, indistinct and not fluted; whole thallus a powdery mass of soredia (pl. 3.1)
                                                             Lepraria incana

49 Thallus thick, pale grey; margin fluted (K+y); fruit bodies black (rare), spores brown with one cross wall (pl. 1.5)   *Diploicia canescens*
–   Thallus thin, yellow-grey to green-grey   50

50 Thallus yellow-grey to grey, often bounded by a dark line so that several adjacent thalli form mosaics on the bark (C+o); black fruit bodies; spores colourless with no cross walls (pl. 2.8)   *Lecidella elaeochroma*
–   Thallus dark green-grey, smooth or cracked (C–) and covered by many small black fruit bodies; spores brown with one cross wall (K.26) (pl. 1.4)   *Buellia punctata*

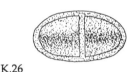

K.26

(Several other lichens, not illustrated, may key out here but differ in chemical and spore characters. To identify these and other rarer lichen species growing on trees, consult Dobson, 1992.)

Chemical tests

Three chemical tests are routinely applied to the surface of lichen thalli. The result of these tests is regarded as positive if a colour is observed (for example, +r indicates a red reaction, +y = yellow, +o = orange, +p = purple); if no colour change occurs, (–) is recorded. In the key above, only positive results are generally noted. Sometimes it is necessary to test the soredia or the medulla (the whitish central part of the thallus exposed by scraping or cutting away the pigmented top layer together with the underlying green algal layer). The results of such tests are recorded as, for example, soredia C+o or medulla K+y.

The three chemical tests involve the use of K (potassium hydroxide) (**corrosive**), C (commercial bleach such as 'Domestos') (**corrosive**) and P (paraphenylenediamine) (**skin irritant and possible carcinogen**). **Proper safety precautions must be used both in carrying out the tests and in making up the solutions.** The three test solutions are made as follows:

C   (commercial domestic bleach such as Domestos) is used undiluted. It should be replaced about once a month as the chlorine is gradually lost from the solution.
K   (potassium hydroxide) is used as a 10% solution in water. This lasts at least three months.
P   (paraphenylenediamine) is made up as a stable solution as follows 1 g paraphenylenediamine, 10 g sodium sulphite, 100 ml water and 2–3 drops of domestic liquid detergent. This solution lasts about three months.

Paraphenylenediamine can be obtained at nominal cost, to personal callers only, from the British Lichen Society (see useful addresses).

A tiny drop of the appropriate solution, C, K or P (or sometimes a drop of K and then a drop of C indicated by KC) is placed on the lichen surface or on the exposed medulla. One way of doing this is to put the three solutions into tiny plastic dropper bottles. Used 'Optrex' eye drop bottles are very convenient **but for obvious reasons they need to be painted in a warning colour (RED) and labelled**. As a further precaution the three dropper bottles should be kept in a screw cap jar again labelled **'POISON'**. When these solutions are used by students, disposable gloves and clear instructions regarding safety and procedures **must be given before any tests are performed.**

Following the application of a drop of the appropriate chemical (K, C or P), the treated lichen surface (or the treated exposed medulla) is observed closely, using a lens if necessary. Any colour changes are recorded. Note that C reactions can fade within about 30 seconds and that P reactions often take a minute or two to develop.

Spore characters

A low power stereomicroscope is required to cut a section of the fruit body and a high power microscope (x100–x400 or more) is needed to observe the spores. Proceed as follows.

For round fruit bodies

Obtain some large sewing needles and a packet of double-sided razor blades used in 'safety razors'. Take a single blade in its paper wrapping and bend it with pliers until it snaps into two usable halves. **Do not take the blade out of its wrapping** or you will certainly cut yourself as the spring steel breaks. It is **essential to instruct students on this point**. (The edge of single sided razor blades is not nearly as sharp and it is difficult to get a good and thin section with these.)

View the round fruit body under the low power stereomicroscope. Using a vertical cut, cut down and push to one side about one third of the fruit body. Then make a second cut parallel to the exposed face of the fruit body and as close as possible to it. Thus one slices a fruit body much as one would a loaf of bread, but on a micro scale. The thin fruit body slice (section) will probably end up resting against the cut face of the fruit body. Using a needle moistened by dipping in water, the section can be picked up and transferred to a drop of water on a microscope slide. A coverslip is placed on top and then the slide is viewed using a high power microscope, first at x100 and then at x400. The spores are contained in small sacs. Tapping the coverslip with the end of the index finger can release spores still in the

sacs. The colour (brown or colourless), the shape (round,
oval or elongate) and the number of cross walls can then be
recorded. Sometimes it is difficult to see the cross walls
because of a mass of oil droplets in the spores. In such cases
a fruit body section is mounted in a drop of 10% potassium
hydroxide (K reagent) which dissolves the oil. Many lichen
asci and spores are very beautiful under the microscope.
Mastering this technique is rewarding. If a micrometer
eyepiece and slide are available the size of the spores can be
determined and this is used for precise determination of
many of the less common lichens (see Dobson, 1992).

For elongate fruit bodies

Moisten the surface of the lichen by covering it with
a film of water. After three or four minutes blot the surface
dry with a paper towel. Then take a large needle and run it
up the slit of the elongate fruit body. The 'jelly' which is
pushed out by the large needle will contain the asci and
spores. If the soaking is too brief, nothing will come out; if it
is too long the jelly will not be easy to pick up. Transfer the
jelly to a drop of water on a microscope slide and view
under a high power microscope at x100 or x400. The sacs
containing the spores are best seen when mounted in water
but mounting in 10% potassium hydroxide is usually
necessary to see the cross walls in the spores.

Explanation of terms used in the identification key

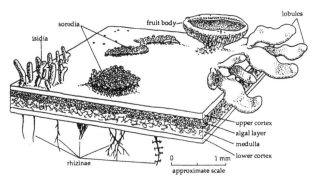

(after Brodo, 1988)

Algal Layer. The layer immediately beneath the upper
protective fungal layer – the upper cortex (see Cortex
below). The algal layer contains either green algae (usually a
species of *Trebouxia*) or cyanobacteria (usually a species of
Nostoc) which photosynthesise and provide the fungal
partner with carbohydrates. When a lichen is torn or cut this
green layer will be seen between the pigmented upper
cortex and the usually white medulla (see Medulla below).

C. Sodium hypochlorite solution containing chlorine which is used for colour tests on lichens. See section on Chemical tests, p. 52.

Cilia. Long hair-like strands coming mainly from the margin of certain species in the genera *Anaptychia*, *Physcia* and *Parmelia*.

Cortex. The outer protective fungal layer which forms the upper and lower surface of the lichen. The upper cortex is often pigmented grey, brown, orange or green and bears the reproductive structures – isidia, soredia or fruit bodies. The lower cortex is usually white, grey, black or brown and often has a number of fungal strands called rhizinae (see below) coming from it and connecting the lichen to the bark (or substratum). Note that crustose and leprose lichens have no lower cortex, the lichen being attached directly to the bark (or substratum).

Foliose lichens have a leaf-like form and a distinct upper and lower surface and are often attached to the bark (or substratum) by fungal strands termed rhizinae (see below). Using a knife it is possible to separate such lichens from the underlying bark.

Fruticose lichens have a bushy or shrubby form, and usually have round or strap-shaped branches which are attached at their base. Such lichens stick up from the bark (or substratum) and can therefore be picked by hand.

Fruit bodies. The round or elongate structures which contain fungal spores in small sacs. Round fruit bodies (apothecia) consist of a central disc which is usually pigmented (red, tan, brown or black). In some lichens the disc is surrounded by a margin that is the same colour as the remainder of the lichen giving the whole structure the appearance of a miniature jam tart. In others this type of margin is lacking so the fruit body looks like a jam tart without the pastry margin.

Isidia. Small peg-, coral- or bun-shaped outgrowths which form in large numbers on the upper surface of some lichens. Isidia may be removed by wind or rainwater and act as a means of vegetative reproduction. If one lands on a suitable surface it can grow into a lichen. Unlike soredia (see below), isidia have a smooth surface. Normally a particular lichen produces either soredia or isidia.

K. Potassium hydroxide solution used for colour tests on lichens (see p. 52).

Leprose lichens consist only of a powdery mass of hyphae and algae producing soredia over the whole surface and having an indistinct margin or edge. Like crustose lichens they can only be removed from the bark (or substratum) on which they grow by removing the substratum as well.

Lobules. Small lobes produced on a few lichens, such as some species of *Peltigera*, which can break off and form a means of vegetative reproduction.

Medulla. The central region of a lichen which is composed of white (occasionally yellowish) fungal strands and probably acts as a storage region. In some lichens the fungal strands in this region have specific chemicals in their walls and these can yield characteristic colour reactions with chemical tests (see p. 52).

P. paraphenylenediamine solution used for colour tests on lichens (see p. 52).

Podetia include the upright-growing branches of lichens in the genus *Cladonia*. These branches are round and hollow and usually attached to a basal cushion of lichen scales. At the tip either red or brown fruit bodies may often be seen.

Rhizinae. The fungal strands which attach the lower surface of many leafy lichens to the underlying bark (substratum). These are often black but may be white, grey or brown.

Soredia. Powdery granules produced on the surface, margin or tip of a lichen. They consist of a few algal cells surrounded by a weft of fungal threads (hyphae) forming tiny rough spheres. Soredia provide one means of vegetative reproduction. They are removed from the lichen by wind and rainwater and can grow into a lichen if they land on a suitable surface.

Spores. These occur in the fruit body and provide lichen fungi with a means of sexual reproduction. Spores occur in small sacs (asci), usually 8 to a sac. Spores may be colourless or brown and both the number of cross walls in each spore and its size may help to identify particular lichens (see p. 53 on sectioning fruit bodies). Under natural conditions spores are forcibly ejected from the sacs and carried away by air currents. If they land on a surface, germinate and meet a suitable alga they can give rise to a new lichen.

Thallus. The name for the whole lichen plant which comprises a fungus and its contained layer of green algae or less often photosynthetic cyanobacteria (formerly called blue-green algae).

12 Techniques and approaches to original work

In schools or colleges, funding and access to analytical equipment are often limited, so it is frequently easier to carry out field studies rather than pollution monitoring studies based on laboratory experiments. The following paragraphs offer guidance and suggestions for both types of project. Readers will glean other useful ideas from the earlier sections, and from the references (p. 66).

School pupils, university students and amateurs have already made important contributions to research on air quality in Britain and Ireland. They have revealed the distribution of melanic forms of moths in Britain, and the changing distribution of *Usnea* in both countries. In a single year, 1971, it also proved possible for the Advisory Centre for Education to organise a nationwide survey of lichens in Britain with the help of 15,000 children. A map of air quality over the whole country was produced as a result of this work (see Richardson, 1975). It is hoped that the present book will stimulate further original studies.

12.1 Distribution surveys

Individuals or groups of students can make important contributions to understanding the pattern of lichen distribution, and can establish air quality zones, around urban or industrial areas. If many people are available, a large quantity of data can be collected in a short time. In Greater Dublin, Ireland, some 2,215 trees were examined and over 5,000 lichen records documented by 104 participating schools during the autumn of 1987 (Ni Lamhna and others, 1988). This survey involved many recorders and required careful preparation. A brief account may help others carrying out similar surveys. First, a small group of coordinators decided on the area to be surveyed, and obtained the necessary maps. One kilometre squares were drawn on these for urban areas and larger squares for outlying rural areas where fewer recorders lived. For each square, a kit was prepared with a small-scale map to locate the square and a blown-up large-scale map of the particular square on which the recorder would accurately indicate the position of each examined tree. Background information, identification chartlets (Dalby, 1981), and a simple key completed each kit which was placed in a separate folder with a grid reference on the outside. After an initial letter inviting participation, individual schools were approached by a team of two: one to see the head teacher and the other to invite the science teacher to become involved. All the teachers who agreed to participate came together for an evening at one school for a briefing session and to obtain pre-allocated kits. [There was some swapping of kits as teachers knew their particular catchment areas.] Each school was given a box of identified lichen specimens that included all the species to be recorded. It was suggested that teachers form recording teams (of two and three students) and should have an initial lichen-study session on trees near the school. It proved necessary to remind many teachers to return the completed data sheets. Such a survey is usually best done in the early autumn before the pressure of end of year exams. Once the coordinators had analysed the results there was a final meeting to report the results to teachers and to a selection of the participating students. Sometimes the results of such surveys prove interesting or controversial as topography and prevailing wind can lead to unexpected patterns of air quality in a city. In such cases the media and local planners may be invited to the meeting. Data may be of relevance to planners considering the siting of new housing estates or industrial activities. Further industry should perhaps be concentrated in areas where the air quality is already poor, leaving housing areas relatively pollution-free. Alternatively, by mixing industry and housing, the whole city might be exposed to an average level of air pollution.

Ideally, lichen distribution surveys should extend from the city centre to rural areas where pollution levels are known to be low. If manpower is limited, the study might be restricted to transects related to the prevailing wind direction, including sites both upwind and downwind of the city centre. If the impact of emissions is expected to be localised, the study may be concentrated in the area of greatest impact, say at sites 100, 200, 400, 800 and 1600 m from the emission source.

The organisers should reconnoitre the survey area to see whether there are enough trees and lichens for study, or whether rock-dwelling lichens might be examined instead. For example, school pupils from Brussels between the ages of 6 and 16 were asked to find specimens of *Lecanora muralis* which is common on stone surfaces in all but the most polluted cities. The pupils then measured the diameter of the largest thallus and recorded the location. As part of pollution survey they established that the mean annual growth rate of this lichen increased from less than 2 to more than 3 mm per year from the centre of the city towards the periphery; and then diminished again in the industrial areas to the northeast of the city. A map of growth rates correlated well with the measured pattern of sulphur dioxide pollution over the city (Sansen & Deronde, 1990). By plotting the distribution over regular time intervals, it has also been possible to show that improvements in air quality enable the lichen to re-colonise highly polluted urban areas (fig. 8) (Henderson-Sellers & Seaward, 1979). In lichen surveys on trees the degree of replication is important. For example, a survey might involve examining ten trees per 1 km square. Certain trees such as conifers (with acid bark) or beech (with very smooth bark) are rejected because they are poor substrata for colonisation by lichens except in the most favoured situations. Usually the commonest one or two deciduous tree species in an area are chosen and the lichens are recorded at a specific height, say 50 cm to 1 m above the ground. The recorders should be instructed to collect data in each square from trees that are widely spaced and which support as good a lichen flora as can be found in that square. It is not useful simply to collect data from the first ten trees encountered, as these may have no lichens on them if they are in unsuitable locations. Suitable trees are usually found along lightly travelled roads, or in parks, churchyards and gardens. The location of sampled trees should be carefully marked on large-scale maps so that they can be re-located. With permission of the owners, the trees may even be marked permanently by means of small numbered aluminium discs fixed by a partially driven-in aluminium nail. This allows for the growth of the tree so that the tag does not become buried. If the tree is felled, it can be sawn up without concern as aluminum tags and nails do not damage machinery in commercial sawmills.

When data recording sheets are prepared, a decision must be made as to how many species are to be recorded. This is done in the light of the taxonomic expertise of those collecting the data. Senior school students in Ireland were asked to identify some twenty lichen species, selected from the different Hawskworth and Rose pollution zones (table 1) (Ni Lamhna and others, 1988). The twenty species are among those illustrated in the plates and for which the identification key has been prepared (pp. 43–52).

The data collected in the field for particular species or groups of species can be hand-plotted on transparent overlays of the base map to reveal the pattern of lichen distribution. Alternatively, a computer may be used with a programme that takes into account the coordinates of each tree in each study square and also records the lichens found. The computer may then be used to plot maps of the distribution of individual species. In the absence of a computer, it is possible to delimit air quality zones by overlaying the distribution of the various species with the help of a light box. This is done with a knowledge of the relative pollution sensitivity of each plotted species using the Hawksworth and Rose scale (table 1). Computer programmes can also be used to produce maps showing the innermost limit of any particular species or group of species. By setting the condition that the innermost limit of a more tolerant species must fall inside one having less tolerance, it is possible to draw air quality maps even when the more tolerant species are not recorded at all sites. The relative tolerance of commonly recorded lichens is shown in table 1.

Another more detailed approach is to calculate Indices of Atmospheric Purity (IAPs) for a study area. A standard number of trees is examined at each site along the supposed pollution gradient. On each tree all epiphytic lichens and mosses are identified and voucher specimens kept. Care should be exercised when making the voucher collections to bear conservation in mind and not to deplete the population of any particular species. This applies

especially to studies in unpolluted areas where rare and
endangered lichens such as *Lobaria* may be found; if in
doubt seek advice from the appropriate lichen society
(p. 74). Each identified species is then assigned a numerical
frequency-coverage value (f) according to the following
scale:

5 – an epiphyte which is very frequent and has a high
    degree of coverage on most trees;
4 – a species which is frequent or has a high degree of
    coverage on some trees;
3 – a species which is infrequent or has a medium degree of
    coverage on some trees;
2 – a species which is very infrequent or has a low degree of
    coverage;
1 – a species which is very rare and has a very low degree of
    coverage.

The IAP values are now calculated for every study site using
the following formula:

$$\text{IAP} = \sum_{N}^{1} \frac{(Q \times f)}{10}$$

In this formula, N is the number of species at a site. Q is the
ecological index of a species determined by adding together
the number of species which occur with that species on trees
at a given site; and then taking the average of the sums for
all the sites where that species was present. The frequency
coverage value, f, is assigned to each species at the field site.
The value $(Q \times f)$ is divided by 10 to yield a smaller more
manageable number.

   The IAP values of all sites are plotted on an outline
map of the study area. Values are usually grouped by range
into IAP zones and up to five zones may be distinguished
(Showman, 1988). IAP zones primarily reflect species
richness, and it therefore helps if, as far as possible, the same
tree species is examined throughout the study area.
Pollution accounts for only about 50% of the IAP variability,
the rest being attributable to habitat differences (Herben &
Liska, 1986).

12.2 Transplant studies

   The transplant technique makes it possible to explore
the effects of different conditions on a standard lichen
material, usually shrubby or leafy lichens. Only common
species, such as *Hypogymnia physodes*, should be selected.
Care should be taken not to deplete the population at a
collection site. Permission must be obtained before
collecting, for example, dead lower branches covered with
this lichen from a forestry plantation. Using methods
outlined in section 8.1, the transplants can be placed at
varying distances around a suspected source of urban or

industrial pollution at sites where permission for the study has been obtained. After two months the transplants will have accumulated near-optimum levels of airborne metals and other pollutants. The effect of an air pollution gradient may be studied by making notes or photographic records of any dying or discoloured lichen lobes at each site. At the end of the transplantation period, electrolyte leakage (see below), ash content (p. 62) or chlorophyll content (see below) can be determined on subsamples. Changes in chlorophyll content or faunal associates can be plotted against the distance of the transplant site from the city or industrial centre. In rural areas it is possible to monitor dust levels or metal contamination by studying lichens growing near quarries or mines. In such situations, lichen transplants may not be needed as it is often possible to use lichens growing naturally on tree trunks or rocks at the various sites.

12.3 Electrolyte leakage and chlorophyll estimation

To study electrolyte leakage, the samples are first placed on moist filter paper in a closed container and allowed to equilibrate in a moist atmosphere for two hours and then rinsed for three seconds. Next, the lichen pieces, 10 to 60 mg, are immersed in vials containing 20 ml of distilled water for five minutes to allow the electrolytes to leak out. The lichen is removed and the conductivity of the water is measured using a specific conductivity meter. The retrieved lichen is dried at 80°C for 24 h and weighed. Results are calculated as conductivity per gram dry weight of lichen per ml of the bathing solution. It is important to let the lichen samples equilibrate in a damp atmosphere in a moist porous pot or on filter paper because dry lichens will leak potassium if wetted rapidly. Lichens are usually collected under dry conditions and it takes them an hour or so to repair the membrane damage which occurs when they dry out in their natural habitat (Alebic-Juretic & Arko-Pijevac, 1989).

Extracting chlorophyll is relatively easy. Marginal lobes or thallus tips (about 15 mg) are placed in 3 ml of dimethyl sulphoxide (DMSO) for 45 minutes at 60°C in the dark. **This chemical should be handled with care.** The extracts are filtered and the absorbance of the extract is then measured in a spectrophotometer at 415 and 435 nm. The amount of degradation of chlorophyll *a* to phaeophytin can be calculated from the optical density ratio 435/415 (Kardish and others, 1987; Belnap & Harper, 1990).

12.4 Metal accumulation

It is possible to locate accumulated metals by using stains which react with specific metals to give coloured products. Thus for lead, small pieces of collected lichen are placed for one hour in sodium rhodizonate stain. This is made by dissolving 50 mg of rhodizonate in 25 ml of tartaric acid buffer (made by dissolving 1.5 g of tartaric acid and 1.9 g of sodium bitartrate in 100 ml of water to give a pH of 2.8).

The lichens are washed, and then thin sections are cut using a sharp razor blade (see p. 53). The sections are examined under a microscope at a magnification of about x400. Red or pink coloration indicates that metal has been accumulated by the lichen and also reveals where in the thallus metal is concentrated (Garty & Thiess, 1990). To check that the stain is effective or to study metal uptake by lichens, thalli from unpolluted areas should be immersed in 20 ppm of lead nitrate for one hour, rinsed in distilled water and then soaked for an hour in the sodium rhodizonate stain. As 'controls', some lichen samples should also be incubated in distilled water instead of lead nitrate solution prior to staining. Again cross-sections of the lichen thalli are cut with a sharp razor blade and observed under the microscope. In a similar way lichens can be stained in a solution of dimethyl glyoxime to test for nickel.

The degree of metal accumulation can be measured by digesting lichen samples in a mixture of concentrated acids and then measuring metal levels in the diluted digest using atomic absorption spectrometry (AA) or by other techniques (p. 24). Sophisticated analytical equipment of this type is seldom available in schools and the digestion procedure is hazardous. An alternative to digestion is ashing the lichen samples, although a proportion of the more volatile metals like mercury, lead and zinc may be lost during the process.

The amount of ash produced from a lichen sample provides an approximate estimate of the degree of accumulation of metal-rich particulates. Samples are collected at varying distances from a pollution source, washed with water to remove surface dust and dried at 80°C for 24 h. Five gram amounts of dried lichen are placed in crucibles. These are heated in a muffle furnace using the following temperature regime: 1 h at 150°C, 2 h at 250°C, 2 h at 350°C and 19 h at 500°C. By raising the temperature slowly, conflagration of the sample and loss of particles in smoke is avoided. After cooling, the ash is weighed and the ash weight per unit weight of lichen plotted against distance from the pollution source.

Metals can be quantified using the techniques, like AA, mentioned above. However, simpler colorimetric methods may also be tried. First, the metals are dissolved out of the ash. This may be done by transferring the ash to a beaker and adding 2 ml of concentrated hydrochloric acid, followed by 2 ml of concentrated nitric acid. This is diluted with distilled water to about 20 ml and transferred via a glass funnel to a 50 or 100 ml volumetric flask. A filter paper in the funnel retains any silica or other insoluble material. The beaker is rinsed out with distilled water which is added to the flask along with enough water to make the required volume. The amount of metal in the extract can be quantified using well described procedures and standard colorimetric test kits (for example, those produced by the Hach Co., available from Philip Harris, Lynn Lane, Shenstone, Staffordshire, UK: Anon., 1988) and where necessary a spectrophotometer. It may be necessary to adjust the pH of the lichen extract with a known amount of alkali, usually sodium hydroxide solution. Lead, zinc, cadmium, copper and iron form coloured compounds with dithizone. These metal dithizones are extractable into various chlorinated organic solvents. Special buffer solutions and other additives are used to make the method specific for a particular metal. After extraction into the organic solvent, the colour is matched immediately with a colour chart to determine the concentration of metal. Procedures and recipes are provided in a recent article in School Science Review (Sanderson & Newton, 1986). The use of acids and organic solvents can be hazardous. Students **must be instructed** on procedures and **must adopt** proper safety precautions such as the use of a fume cupboard, protective clothing and safety glasses. Furthermore, the safe disposal of wastes has to be considered.

Although these colorimetric techniques are not as sensitive as atomic absorption spectrometry, it should be possible to quantify the levels of the most abundant metals in lichen samples collected at varying distances from a pollution source. The following regression model can then be used to determine the relationship between the concentration and distance:

$$C(d) = md^{-n} + b$$

where C denotes the elemental concentration (usually in units of $\mu g\ g^{-1}$) at the distance d (in units of m or km), m is the slope of the regression line, n is the distance power coefficient, and b is the intercept. In using this model, $\log [C(d) - b']$ is plotted against $\log d$ where b' corresponds to the elemental level recorded furthest from the pollution source. The value of n is then assessed from the plot and rounded off to the nearest whole number. Regression analysis of the plot of $C(d)$ versus d^{-n} evaluates b, the background level. From these plots the distance of metal fallout around the industrial or urban pollution source may be established.

12.5 Microfaunal associates

The diversity and abundance of the microfauna associated with lichens can be altered by air pollution (p. 13). The basic procedure for assessing any effects is not difficult. Lichen samples of about 2 x 2 cm are incubated in 5 ml of sterile water in small Petri dishes for 4–9 days at room temperature. The samples and the water are then examined under a good stereoscopic microscope. Detailed study of the smaller creatures involves transferring a drop of the water with a Pasteur pipette to a microscope slide. A coverslip is placed gently in position and then the slide is examined under a compound microscope at about x400 magnification. The animals can be identified with the aid of Hingley (1992), Fitter & Manuel (1986) and Streble & Krauter (1973). The book by Morgan & King (1976) should be consulted for tardigrade identification and the publications of Curds (1982) and Curds and others (1983) used for the ciliated protozoa.

12.6 Melanic moths

The effects of changing patterns of air pollution can be revealed by examining the ratio of melanic to typical forms of moth species which exhibit lichen crypsis (p. 40). Data collected by Open University students during 1983 and 1984 proved invaluable for following this change in the peppered moth. There are many other moths with melanic forms. A well-placed mercury vapour moth trap is likely to yield valuable information at almost any time of year. In the last three months of the year, the Brindled Crescent and the highly variable Mottled Umber readily come to such traps. Even in midwinter, several moths which have melanic forms are on the wing, including the Pale Brindled Beauty and the Dotted Border in January to March. These are closely followed by the Spring Usher, the Brindled Beauty and the Early Thorn (Majerus, 1989). Even more valuable information can be collected if there is cooperation between a series of schools or colleges. Several traps can then be run concurrently at different locations at varying distances from a conurbation, or in a range of contrasting habitats such as deciduous and coniferous woodlands (Majerus, 1989). Running a moth trap and examining the catch can be instructive and there are several well-illustrated moth identification manuals, for example, South (1939) and Skinner (1984).

12.7 Preparation of reports

In any pollution monitoring study a report has eventually to be prepared. Writing up is an important part of a research project. The report should outline the history of the urban or industrial area, describe the methods used and give the results of the lichen survey or experiments undertaken. A discussion should follow which relates results to any available data from mechanical monitors, information on pollution sources, and the effects of prevailing wind and weather. Photographs and other illustrations greatly improve the visual impact of the report which should also, where appropriate, include an overall map of air quality as defined by the distribution of particular lichens whose sensitivity is well established. If possible, multiple copies should be prepared so that the original information obtained can be distributed to other workers.

12.8 Publication of results

A really thorough, critical investigation that has established new information of general interest may be worth publishing. Guidance on the publication of research project results is given by Majerus & Kearns (1989). Journals that might consider short papers on lichens and pollution include *British Lichen Society Bulletin*, *The Lichenologist*, *The Bryologist*, *International Lichenological Newsletter*, *School Science Review* and *Journal of Biological Education*. Examine recent numbers of these journals to see what sort of thing they publish, and then write a paper along similar lines, keeping it as short and concise as possible while still presenting enough information to enable the study to be repeated and to establish the conclusions. Advice from an expert is helpful at this stage.

It is an unbreakable convention of scientific publication that results are reported with scrupulous accuracy and honesty. It is therefore essential to keep detailed records throughout the investigation, and to distinguish in the write-up between certainty and probability, and between deduction and speculation.

It will often be necessary to apply statistical techniques to test the significance of the findings. A book such as the Open University Project Guide (Chalmers & Parker, 1989) will help, but expert advice can contribute much to the planning, as well as to the analysis, of the work.

References and further reading

The reading list below is divided into two parts. The first comprises a list of books on biological monitors and other items referred to in the Author's Preface. These will provide valuable background information. The second list includes the references (books and research papers) cited in the text by author and date. Some of the books and especially the journals will not be available at local school or public libraries. Often a nearby university library may possess the required book or journal and a visit can usually be arranged to see or obtain a photocopy of the item of interest. Alternatively, your local Public Library can borrow or photocopy a work for you via the British Library Document Supply Centre. This may take several weeks. It is important to present your librarian with the complete reference, exactly as listed below. Without all details, it will be difficult for the libararian to find the item and it will at the very least involve extra delay.

Books on biological monitors and items cited in preface

Burton, A. (1986). *Biological Monitors of Environmental Contamination (Plants)*. [MARC Report No. 32.] London: King's College Monitoring Assessment Centre.

Hawksworth, D.L. & Rose, F. (1976). *Lichens as Pollution Monitors*. Studies in Biology No. 66. London: Edward Arnold.

Henderson, A. (1990). Literature on air pollution and Lichens XXXII. *Lichenologist* **22**,397–404.

Jeffrey, D.W. & Madden, B. (1991). *Bioindicators and Environmental Management*. London: Academic Press.

Manning, W.J. & Feder, W.A. (1980). *Biomonitoring Air Pollutants with Plants*. London: Applied Science.

Nash, T.H. & Wirth, V. (ed.) (1988). *Lichens, Bryophytes and Air Quality*. Berlin: J. Cramer.

Richardson, D.H.S. (1987). *Biological Indicators of Pollution*. Dublin: Royal Irish Academy

Text references

Aberg, B. & Hungate, F.P. (1967). *Radioecological Concentration Processes*. Oxford: Pergamon Press.

Alebic-Juretic, A. & Arko-Pijevac, M. (1989). Air pollution damage to cell membranes in lichens – results of simple biological tests applied in Rijeka. *Water, Air and Soil Pollution* **47**, 25–33.

Andre, H.M., Bolly, C. & Lebrun, P. (1982). Monitoring and mapping air pollution through an animal indicator. A new and quick method. *Journal of Applied Ecology* **19**, 107–112.

Anonymous (1988). *Plant Tissue and Sap Analysis Manual*. Hach Company Publication, No. 3118, Loveland, Colorado, 252 pp.

Bacci, E., Calamari, D., Gaggi, C., Fanelli, R., Focardi, S. & Morosini, M. (1986). Chlorinated hydrocarbons in lichen and moss samples from the Antarctic Peninsula. *Chemosphere* **15**, 747–754.

Baldi, F. (1988). Mercury pollution in the soil and mosses around a geothermal plant. *Water, Air and Soil Pollution* **38**, 111–19.

Bargagli, R., Iosco, F.P. & D'Amato, M.L. (1987a). Zonation of trace metal accumulation in three species of epiphytic lichens belonging to the genus *Parmelia*. *Cryptogamie Bryologie et Lichenologie* **8**, 331–7.

Bargagli, R., Iosco, F.P. & Barghigiani, C. (1987b). Assessment of mercury disposal in an abandoned mining area by soil and lichens. *Water, Air and Soil Pollution* **36**, 219–25.

Bates, J.W., Bell, J.N.B. & Farmer, A.M. (1990). Epiphytic recolonization of oaks along a gradient of air pollution in south-east England. *Environmental Pollution* **68**, 81–99.

Beck, J.N. & Ramelow, G.J. (1990). Use of lichen biomass to monitor dissolved metals in natural waters. *Bulletin of Environmental Contamination and Toxicology* **44**, 302–8.

Beckett, P.J., Boileau, L.J.R., Padovan, D. & Richardson, D.H.S. (1982). Lichens and mosses as monitors of industrial activity associated with uranium mining in Northern Ontario, Canada – Part 2: Distance dependent uranium and lead accumulation. *Environmental Pollution (Series B)* **4**, 91–107.

Belnap, J. & Harper, K.T. (1990). Effects of a coal-fired power plant on the rock lichen *Rhizoplaca melanophthalma*: chlorophyll degradation and electrolyte leakage. *Bryologist* **93**, 309–12.

Bishop, J.A. & Cook, L.M. (1975). Moths, melanism and clean air. *Scientific American* **232**, 90–9.

Boonpragob, K. & Nash, T.H. (1990). Seasonal variation of elemental status in the lichen *Ramalina menzeisii* Tayl. from two sites in southern California: Evidence for dry deposition accumulation. *Environmental and Experimental Botany* **30**, 415–28.

Boonpragob, K. & Nash, T.H. (1991). Physiological responses of the lichen *Ramalina menzeisii* Tayl. to the Los Angeles urban environment. *Environmental and Experimental Botany* (in press).

Boonpragob, K., Nash, T.H., & Fox, C.A. (1989). Seasonal deposition patterns of acidic ions and ammonium to the lichen *Ramalina menziesii* Tayl. in southern California. *Environmental and Experimental Botany* **29**, 187–97.

Brodo, I.M. (1961). Transplant experiments with corticolous lichens using a new technique. *Ecology* **42**, 838–41.

Brodo, I.M. (1968). *The Lichens of Long Island, New York; A Vegetational and Floristic Analysis*. New York State Museum and Science Service Bulletin 410, 1–330.

Brodo, I.M. (1988). *Lichens of the Ottawa Region*, 2nd edn. Ottawa Field Naturalists' Club special publication No. 3, Ottawa.

Brown, D.H. (ed.) (1985). *Recent Advances in Lichen Physiology*. London: Plenum Press.

Case, J.W. & Krouse, H.R. (1980). Variations in sulphur content and stable sulphur isotope composition in vegetation near an SO_2 source at Fox Creek, Alberta, Canada. *Oecologia (Berlin)* **44**, 248–57.

Chalmers, N. & Parker, P. (1989). *The OU Project Guide. Fieldwork and statistics for ecological projects*, 2nd edn. Field Studies Council Occasional Publications No. 9.

Connor, M., Dempsey, E., Smyth, M.R. & Richardson, D.H.S. (1991). Determination of some metal ions using lichen-modified carbon paste electrodes. *Electroanalysis* **3**, 331–6.

Cook, L.M., Rigby, K.D. & Seaward, M.R.D. (1990). Melanic moths and changes in epiphytic vegetation in north-west England and north Wales. *Biological Journal of the Linnean Society* **39**, 343–54.

Coppins, B.J. & Lambley, P.W. (1974). Changes in the lichen flora of the parish of Mendlesham, Suffolk, during the last fifty years. *Suffolk Natural History* **16**, 319–35.

Curds, C.R. (1982). *British and Other Freshwater Ciliated Protozoa*. Part I. Synopses of the British Fauna (New Series) No. 22. Cambridge: Cambridge University Press.

Curds, C.R., Gates, M.A. & Roberts, D.M.L. (1983). *British and Other Freshwater Ciliated Protozoa*. Part II. Synopsis of the British Fauna (New Series) No. 23. Cambridge: Cambridge University Press.

Dalby, C. (1981). *Lichens and Air Pollution*. Wallchart and A4-sized chartlets. London: British Museum (Natural History).

Dalby, C. (1987). *Lichens on Rocky Seashores*. Wallchart and A4-sized chartlets. London: British Museum (Natural History).

Davies, F.B.M. & Notcutt, G. (1988). Accumulation of fluoride by lichens in the vicinity of Etna volcano. *Water, Air and Soil Pollution* **42**, 365–71.

Davies, F.B.M. & Notcutt, G. (1989). Accumulation of volcanogenic fluorides by lichens. *Fluoride* **22**, 59–65.

De Bakker, A.J. (1989). Effects of ammonia emission on epiphytic lichen vegetation. *Acta Botanica Neerlandica* **38**, 337–42.

De Bruin, M., Van Wijk, P., Van Assema, R. & De Roos, C. (1986). *The Use of Multi-element Concentration Data Sets Obtained by I.N.A.A. in the Identification of Sources of Environmental Pollutants.* Interuniversitair Reactor Institute Report, Delft.

Dobson, F.S. (1992). *Lichens, an Illustrated Guide*, 3rd edn. Slough: Richmond Publishing.

Farkas, E., Lokos, L. & Verseghy, K. (1985). Lichens as indicators of air pollution in the Budapest agglomeration. *Acta Botanica Hungarica* **31**, 45–68.

Fenton, A.F. (1960). Lichens as indicators of atmospheric pollution. *Irish Naturalists' Journal* **13**, 153–9.

Fitter, R. & Manuel, R. (1986) *Field Guide to the Freshwater Life of Britain and North-West Europe.* London: Collins.

Gailey, F.A.Y. & Lloyd, O.L. (1986). Methodological investigations into low technology monitoring of atmospheric metal pollution: Part 2 – The effects of length of exposure on metal concentrations. *Environmental Pollution* **12**, 61–74.

Gailey, F.A.Y., Smith, G.H., Rintoul, L.J. & Lloyd, O.L. (1985). Metal deposition patterns in central Scotland as determined by lichen transplants. *Environmental Monitoring and Assessment* **5**, 291–309.

Garty, J. (1988). Comparisons between the metal content of transplanted lichen before and after the start-up of a coal-fired power station in Israel. *Canadian Journal of Botany* **66**, 668–71.

Garty, J. & Hagemeyer, J. (1988). Heavy metals in the lichen *Ramalina duriaei* transplanted at biomonitoring stations in the region of coal-fired power plant in Israel after 3 years operation. *Water, Air and Soil Pollution* **38**, 311–23.

Garty, J. & Theiss, H. (1990). The localization of lead in the lichen *Ramalina duriaei* (de Not.) Bagl. *Botanica Acta* **103**, 311–14.

Gilbert, O.L. (1968). *Biological Indicators of Air Pollution.* PhD Thesis, University of Newcastle upon Tyne.

Gilbert, O.L. (1971). Some indirect effects of air pollution on bark-living invertebrates. *Journal of Applied Ecology* **8**, 77–84.

Gilbert, O.L. (1976). An alkaline dust effect on epiphytic lichens. *Lichenologist* **8**, 173–8.

Gilbert, O.L. (1990). A successful transplant operation involving *Lobaria amplissima*. *Lichenologist* **23**, 73–77.

Gouaux, P. & Vincent, J.P. (1990). Evidence and control of pollution action on lichens (*Peltigera canina*) using infrared colour film. *Science of the Total Environment* **95**, 181–1.

Goyal, R. & Seaward, M.R.D. (1982). Metal uptake in terricolous lichens. III. Translocation in the thallus of *Peltigera canina. New Phytologist* **90**, 85–98.

Grindon, L.H. (1859). *The Manchester Flora.* London: W. White.

Hallinback, T. (1989). Occurrence and ecology of the lichen *Lobaria scrobiculata* in Southern Sweden. *Lichenologist* **21**, 331–42.

Hawksworth, D.L. (1990). Long-term effects of air pollutants on lichen communities in Europe and North America. In *The Earth in Transition: Patterns and Processes of Biotic Impoverishment*, ed. G.M. Woodwell, pp. 45–64. Cambridge: Cambridge University Press .

Hawksworth, D.L. & Ahti, T. (1990). A bibliographic guide to the lichen floras of the world. *Lichenologist* **22**, 1–78.

Hawksworth, D.L. & McManus, P.M. (1989). Lichen recolonization in London (UK) under conditions of rapidly falling sulphur dioxide levels and the concept of zone skipping. *Botanical Journal of the Linnean Society* **100**, 99–110.

Hawksworth, D.L. & Rose, F. (1970). Qualitative scale for estimating sulphur dioxide air pollution in England and Wales using epiphytic lichens. *Nature (London)* **227**, 145–8.

Hawksworth, D.L. & Rose, F. (1976). *Lichens as Pollution Monitors.* Studies in Biology No. 66. London: Edward Arnold.

Henderson-Sellers, A. & Seaward, M.R.D. (1979). Monitoring lichen re-invasion of ameliorating environments. *Environmental Pollution* 19, 207–13.

Herben, T. & Liska, J. (1986). A simulation study of the effect of flora composition, study design and index choice on the predictive power of lichen indication. *Lichenologist* 18, 349–62.

Hingley, M. (1992). Microscopic animals in *Sphagnum*. Naturalists' Handbooks, Richmond Publishing, Slough (in preparation).

Hohenemser, C., Deicher, M., Hofasss, H., Lindnep, G., Recknagel, E. and Budnick, J.I. (1986). Agricultural impact of Chernobyl: a warning. *Nature (London)* 321, 817.

Holleman, D.F., Luick, J.R. & R.G. White (1979). Lichen intake estimates for reindeer and caribou during winter. *Journal of Wildlife Management* 43, 192–201.

Holm, E. (1977). *Plutonium Isotopes in the Environment.* Thesis, University of Lund, Lund.

Holm, E. & Persson, R.B.R. (1978). Biophysical aspects of Am-241 and Pu-241 in the environment. *Radiation and Environmental Biophysics* 15, 261–76.

Holopainen, T.H. (1984a). Types and distribution of ultrastructural symptoms in epiphytic lichens in several urban and industrial environments in Finland. *Annales Botanici Fennici* 21, 213–29.

Holopainen, T.H. (1984b). Cellular injuries in epiphytic lichens transplanted to air polluted areas. *Nordic Journal of Botany* 4, 393–408.

Holopainen, T. & Kauppi, M. (1989). A comparison of light, fluorescence and electron microscopic observations in assessing the SO_2 injury of lichens under different moisture conditons. *Lichenologist* 21, 119–34.

Jahns, H.M. (1983). *Collins Guide to Ferns, Mosses and Lichens of Britain and North and Central Europe.* London: Collins.

Jones, D. (1988). Lichens and pedogenesis. In *Handbook of Lichenology* Vol. III, ed. M. Galun, pp. 109–24. Boca Raton: CRC Press .

Kardish, N., Ronen, R., Bubrick, P. & Garty, J. (1987). The influence of air pollution on the concentration of ATP and on chlorophyll degradation in the lichen *Ramalina duriaei* (De Not.) Bagl. *New Phytologist* 106, 697–706.

Kershaw, K.A. (1985). *Physiological Ecology of Lichens.* Cambridge: Cambridge University Press.

Kettlewell, H.B.D. (1959). Darwin's missing evidence. *Scientific American* (March 1959).

Kettlewell, H.B.D. (1973). *The Evolution of Melanism.* Oxford: Clarendon Press.

Kirkey, S. (1990). Nursing mothers urged not to panic over dioxins report. *Ottawa Citizen* (30 October 1990), pp. 1–2.

Kral, R., Kryzova, L. & Liska, J. (1989). Background concentrations of lead and cadmium in the lichen *Hypogymnia physodes. Science of the Total Environment* 84, 201–9.

Krouse, H.R. (1977). Sulphur isotope abundance elucidates uptake of atmospheric sulphur emissions by vegetation. *Nature (London)* 265, 45–6.

Kwapulinski, J., Seaward, M.R.D. & Bylinska, E.A. (1985). Uptake of 226 radium and 228 radium by the lichen genus *Umbilicaria. Science of the Total Environment* 41, 135–41.

Laaksovirta, K., Olkkonen, H. & Alakuijala, P. (1976). Observations on the lead content of lichen and bark adjacent to a highway in southern Finland. *Environmental Pollution* 11, 247–55.

Lange, O.L., Heber, U., Schulze, E.D. and Ziegler, H. (1989). Atmospheric pollutants and plant metabolism. *Ecological Studies* 7, 238–73.

Laundon, J.R. (1986). *Lichens.* Aylesbury: Shire Publications.

Lawrey, J.D. (1984). *Biology of Lichenized Fungi.* pp. 232–47. New York: Praeger.

Lawrey, J.D. & Hale, M.E. (1988). Lichen evidence for changes in atmospheric pollution in Shenandoah National Park, Virginia. *Bryologist* 91, 21–3.

LeBlanc, F., Rao, D.N. & Comeau, G. (1972). Indices of atmospheric purity and fluoride pollution pattern in Arvida, Quebec. *Canadian Journal of Botany* 50, 991–8.

LeBlanc, F. & De Sloover, J. (1970). Relation between industrialization and the distribution and growth of epiphytic lichens and mosses in Montreal. *Canadian Journal of Botany* **48**, 1485–96.

Lechowicz, M.J. (1987). Resistance of the caribou lichen *Cladina stellaris* (Opiz.) Brodo to growth reductions by simulated acid rain. *Water, Air and Soil Pollution* **34**, 71–7.

Lerond, M. (1978). Courbes d'isopollution de la région de Rouen obtenues par l'observation des lichenes epiphytiques. *Bulletin de la Société Linné Normandie* **106**, 73–84.

Liden, K. & Gustafsson, M. (1967). Relationship and seasonal variation of ^{137}Cs in lichen, reindeer and man in northern Sweden, 1961–1965. In *Radioecological Concentration Processes*, ed. B. Aberg & F.P. Hungate, pp. 193–208. Oxford: Pergamon Press.

Liebert, T.G. & Brakefield, P.M. (1987). Behavioural studies on the peppered moth *Biston betularia* and a discussion of the role of pollution and lichens in industrial melanism. *Biological Journal of the Linnean Society* **31**, 129–50.

Looney, J.H.H., Webber, C.E., Nieboer, E., Stetsko, P.I. & Kershaw, K.A. (1986). Interrelationships between concentrations of ^{137}Cs and various stable elements in three lichen species. *Health Physics* **50**, 148–52.

Luhmann, H.J., Wietschorke, G. & Kreeb, K. H. (1989). Influences of combined temperature, lead and herbicide stresses on lichen fluorescence and their mathematical modelling. *Photosynthetica* **23**, 71–6.

Mackenzie, D. (1986). The rad-dosed reindeer. *New Scientist* No. 1539, 37–40.

McPhee, J. (1989). *The Control of Nature*. Toronto: Harper-Collins.

Majerus, M.E.N. (1989). Melanic polymorphism in the peppered moth, *Biston betularia*, and other Lepidoptera. *Journal of Biologial Education* **23**, 267–84.

Majerus, M.E.N. & Kearns, P.W.E. (1989). *Ladybirds*. Naturalists' Handbooks No. 10, Slough: Richmond Publishing.

Morgan, C.I. & King, P.E. (1976). *British Tardigrades*. Synopses of the British Fauna (New Series) No. 9. London: Academic Press.

Nash, T.H. (1989). Metal tolerance in lichens. In *Heavy Metal Tolerance in Plants: Evolutionary Aspects*, ed. J. Shaw, pp. 119–32. Boca Raton: CRC Press.

Nieboer, E. & Richardson, D.H.S. (1980). The replacement of the nondescript term 'heavy metal' by a biologically and chemically significant classification of metal ions. *Environmental Pollution (Series B)* **1**, 3–26.

Nieboer, E. & Richardson, D.H.S. (1981). Lichens as monitors of atmospheric deposition. In *Atmospheric Pollutants in Natural Waters*, ed. S.J. Eisenreich, pp. 339–88. Ann Arbor: Ann Arbor Science.

Nieboer, E., McFarlane, J.D. & Richardson, D.H.S. (1984). Modifications of plant cell buffering capacities by gaseous air pollutants. In *Gaseous Air Pollutants and Plant Metabolism*, ed. M. Koziol & F.R. Whatley, pp. 313–30. London: Butterworths.

Ni Lamhna, E., Richardson, D.H.S., Dowding, P. & Ni Grainne, E. (1988). *An Air Quality Survey of the Greater Dublin Area carried out by Second Level Students*. Dublin: An Foras Forbartha.

Nimis, P.L., Castello, M. & Perotti, M. (1990). Lichens as biomonitors of sulphur dioxide pollution in La Spezia (Northern Italy). *Lichenologist* **22**, 333–44.

Nylander, W. (1866). Les lichens du Jardin du Luxembourg. *Bulletin de la Société de Botanique de France* **13**, 364–72.

Nylander, W. (1896). *Les lichens des Environs de Paris*. Paris.

O'Clery, C. (1986). The Chernobyl fallout: How Lapland is paying the price. *Irish Times* (weekend supplement), 13 September.

O'Leary, D. (1988). *Air Quality in the National Parks*. Natural Resources Report 88-1, US National Park Service, Denver.

Papastefanou, C., Manolopoulou, M. & Sawidis, T. (1989). Lichens and mosses: Biological monitors of radioactive fallout from Chernobyl reactor accident. *Journal of Environmental Radioactivity* **9**, 19–207.

Perkins, D.F. (1980). Accumulation and effects of airborne fluoride on the saxicolous lichen *Ramalina siliquosa*. *1979 Annual Report, Institute of Terrestrial Ecology*, pp. 81–4. Cambridge: ITE.

Perkins, D.F. & Millar, R.O. (1987a). Effects of airborne fluoride emissions near an aluminium works in Wales: Part 1 – Corticolous lichens growing on broadleaved trees. *Environmental Pollution* 47, 63–78.

Perkins, D.F. & Millar, R.O. (1987b). Effects of airborne fluoride emissions near an aluminium works in Wales: Part 2 – Saxicolous lichens growing on rocks and walls. *Environmental Pollution* 48, 185–96.

Persson, B. (1967). ^{55}Fe from fallout in lichen, reindeer and lapps. In *Radioecological Concentration Processes*, ed. B. Aberg & F.P. Hungate, pp. 253–66. Oxford: Pergamon Press.

Pilegaard, K. (1978). Airborne metals and SO_2 monitored by epiphytic lichens in an industrial area. *Environmental Pollution* 17, 81–92.

Pilegaard, K., Rasmussen, L. & Gydesen, H. (1979). Atmospheric background deposition of heavy metals in Denmark monitored by epiphytic cryptogams. *Journal of Applied Ecology* 16, 843–53.

Pruit, W.O. (1963). Lichen, caribou and high radiation in Eskimos. *Audubon Magazine* 65, 284–7.

Puckett, K.K. & Finegan, E.J. (1980). An analysis of the element content of lichens from the Northwest Territories, Canada. *Canadian Journal of Botany* 58, 2073–89.

Richardson, D.H.S. (1975). *The Vanishing Lichens: Their History, Biology and Importance*. Newton Abbot: David and Charles.

Richardson, D.H.S. (1988). Understanding the pollution sensitivity of lichens. *Botanical Journal of the Linnean Society* 96, 31–43.

Richardson, D.H.S. (1991). Lichens and Man. In *Frontiers in Mycology*, ed. D.L. Hawksworth, pp. 187–210. London: International Mycological Institute.

Richardson, D.H.S. & Nieboer, E. (1981). Lichens and pollution monitoring. *Endeavour (New Series)* 5, 127–33.

Richardson, D.H.S. & Nieboer, E. (1983). The uptake of nickel ions by lichen thalli of the genera *Umbilicaria* and *Peltigera*. *Lichenologist* 15, 8–88.

Richardson, D.H.S., Kiang, S., Ahmadjian, V. & Nieboer, E. (1985). Lead and uranium uptake by lichens. In *Lichen Physiololgy and Cell Biology*, ed. D.H. Brown, pp. 227–46 London: Plenum Publishing.

Richardson, R.M. (1982). Blood-lead concentrations in three to eight year old schoolchildren from Dublin City and rural County Wicklow. *Irish Journal of Medical Science* 151, 203–10.

Roberts, D. & Zimmer, D. (1990). Microfaunal communities associated with epiphytic lichens in Belfast. *Lichenologist* 22, 163–72.

Roy-Arcand, L., Delisle, C.E. & Briere, F.G. (1989). Effects of simulated acid precipitation on the metabolic activity of *Cladina stellaris*. *Canadian Journal of Botany* 67, 1796–1802.

Saeki, M., Kunii, K., Seki, T., Sugiyama, K., Suzuki, T. & Shishido, S. (1977). Metal burdens in urban lichens. *Environmental Research* 13, 256–66.

Sanderson, P.L. & Newton, G. (1986). The pollution detectives. *School Science Review* 68, 224–35.

Sansen, U. & Deronde, L. (1990). Lichenometry of *Lecanora muralis* as a method for an air pollution survey by school children. *Mémoires de la Société Royale de Botanique de Belgique* 12, 100–10.

Schonbeck, H. & van Hut, H. (1971). Exposure of lichens for the recognition and the evaluation of air pollutants. In *Identification and Measurement of Environmental Pollutants*, Proceedings of the International Symposium, Ottawa, Ontario, June 1971, pp. 329–34. Ottawa: National Research Council of Canada.

Schwartzman, D., Kasim, M., Stieff, L. & Johnston, J.H. (1987). Quantitative monitoring of airborne lead pollution by foliose lichens. *Water, Air and Soil Pollution* 32, 363–78.

Schwartzman, D.W., Stieff, L., Kasim, M., Kombe, E., Aung, S., Atekwana, E., Johnson, J. & Schwartzman, K. (1991). An ion-exchange model of lead-210 and lead uptake in a foliose lichen; application to quantitative monitoring of airborne lead fallout. *Science of the Total Environment* 100, 319–36.

Scott, M.G. , Hutchinson, T.C. & Feth, M.J. (1989). A comparison of the effects on Canadian boreal forest lichens of nitric and sulphuric acid as sources of rain acidity. *New Phytologist* 111, 663–71.

Seaward, M.R.D. (1987). Effects of quantitative and qualitative changes in air pollution on the ecological and geographical performance of lichens. In *The Effects of Atmospheric Pollutants on Forests, Wetlands and Agricultural Ecosystems*, ed. T. Hutchinson, pp. 439–50. Berlin: Springer-Verlag.

Seaward, M.R.D. (1989). Lichens as monitors of recent changes in air pollution. *Plants Today* 2, 64–9.

Seaward, M.R.D. (1990). The lichen flora of industrial Teesside. *Naturalist* 115, 73–9.

Seaward, M.R.D. & Letrouit-Galinou, M.-A. (1991). Lichens return to the Jardin du Luxemburg after an absence of almost a century. *Lichenologist* 23, 181–6.

Seaward, M.R.D., Heslop, J.A., Green, D. & Bylinska, E.A. (1988). Recent levels of radionuclides in lichens from southwest Poland with particular reference to [134]Cs and [137]Cs. *Journal of Environmental Radioactivity* 7, 123–9.

Seyd, E. & Seaward, M.R.D. (1984). The association of oribatid mites with lichens. *Zoological Journal of the Linnean Society* 8, 369–420.

Shaw, J. (ed.) (1990). *Evolutionary Aspects of Heavy Metal Tolerance in Plants*. Boca Raton: CRC Press.

Showman, R.E. (1988). Mapping air quality with lichens, the North American experience. In *Lichens, Bryophytes and Air Quality*, ed. T.H. Nash & V. Wirth, pp. 67–89. Bibliotheca Lichenologica, Vol. 30. Berlin: J. Cramer.

Sigal, L.L. & Nash, T.H. (1983). Lichen communities on conifers in southern California mountains: an ecological survey relative to oxidant air pollution. *Ecology* 64, 1343–54.

Skinner, B. (1984). *Colour Identification Guide to Moths of the British Isles* (Macrolepidoptera). Harmondsworth: Viking.

Skye, E. (1958). Luftforeningars inverkan pa pusk- och bladlavfloran kring skifferoljeverket i Narakes Kvarntorp. *Svensk botanische Tidtscrift* 52, 133–90.

Slack, N.G. (1988). The ecological importance of lichens and bryophytes. In *Lichens, Bryophytes and Air Quality*, ed. T.H. Nash & V. Wirth, pp. 23–54. Bibliotheca Lichenologica vol. 30. Berlin: J. Cramer.

Sloof, J.E. & Wolterbeek, H.T. (1991). National trace-element air pollution monitoring survey using epiphytic lichens. *Lichenologist* 23, 139–66.

Smith, F.B. & Clark, M.J. (1986). Radionuclide deposition from the Chernobyl cloud. *Nature (London)* 332, 690–1.

Smith, F.B. & Clark, M.J. (1989). *The transport and deposition of airborne debris from the Chernobyl nuclear power plant accident with special emphasis on the consequences to the United Kingdom*. Meteorological Office Scientific Paper No. 42. London: HMSO, 56p.

Smith, J.N. & Ellis, K.M. (1990). Time dependent transport of Chernobyl radioactivity between atmospheric and lichen phases in eastern Canada. *Journal of Environmental Radioactivity* 11, 151–68.

Sochting, U. (1987). Injured reindeer lichens in Danish lichen heaths. *Graphis Scripta* 1, 103–6.

Sochting, U. (1990). Reindeer lichens injured in Denmark. *British Lichen Society Bulletin* No. 67, 1–4.

South, R. (1939). *The Moths of the British Isles*, 3rd edn. London: Warne.

Starling, A. P. & Ross, I.S. (1990). Uptake of manganese by *Penicillium notatum*. *Microbios* 63, 93–100.

Streble, H. & Krauter, D. (1973). *Das Leben im Wassertropfen*. Stuttgart: Kosmos, Franckhsche Verlagshandlung.

Takala, K., Olkkonen, H., Ikonen, J., Jaasekelainen, J. & Puumalainen, P. (1985). Total sulphur contents of epiphytic and terricolous lichens in Finland. *Annales Botanici Fennici* 22, 91–100.

Takala, K., Olkkonen, H., Jaaskelainen, J., & Selkainaho, K. (1990). The total chlorine content of epiphytic and terricolous lichens and birch bark in Finland. *Annales Botanici Fennici* 27, 131–7.

Takala, K., Olkkonen, H. & Krouse, H.R. (1991). Sulphur isotope composition of epiphytic and terricolous lichens and pine bark in Finland. *Environmental Pollution* 69, 337–48.

Taylor, H.W., Hutchinson, E.A., McInnes, K.L. & Svoboda, J. (1979). Cosmos 954: Search for airborne radioactivity in lichens in the crash area, Northwest Territories, Canada. *Science* 205, 1383–5.

Thompson, R.L., Ramelow, G.J., Beck, J.N., Langley, M.P., Young, J.C. & Casserly, D.M. (1987). A study of airborne metals in Calcasieu Parish using the lichens *Parmelia praesorediosa* and *Ramalina stenospora*. *Water, Air and Soil Pollution* 36, 295–309.

Tomassini, F.D.K., Puckett, K.J., Nieboer, E. & Richardson, D.H.S. (1976). Determination of copper, iron, nickel and sulphur by X-ray fluorescence analysis in lichens from the Mackenzie Valley, Northwest Territories, and the Sudbury District, Ontario. *Canadian Journal of Botany* 54, 1591–1603.

Turk, R. (1988). Bioindikation von Luftverundreingdungen mittels Flechten. In *Okophysiologische Probleme durch Luftverunreinigungen*, ed. D. Grill & H. Guttenberger, pp. 13–27. Graz: Karl-Franzens-Universität.

Turner, D. & Borrer, W. (1839). *Specimen of a Lichenographia Britannica*. Yarmouth, privately printed.

Tyler, G. (1989). Uptake, retention and toxicity of heavy metals in lichens. *Water, Air and Soil Pollution* 47, 321–33.

Van Assche, F. & Clijsters, H. (1990). Effects of metals on enzyme activity in plants. *Plant, Cell and Environment* 13, 19–206.

Vestergaard, N.K., Stephansen, U., Rasmussen, L. & Pilegaard, K. (1986). Airborne heavy metal pollution in the environment of a Danish steel plant. *Water, Air and Soil Pollution* 27, 363–77.

Villeneuve, J.P. & Holm, E. (1984). Atmospheric background of chlorinated hydrocarbons studied in Swedish lichens. *Chemosphere* 13, 1133–38.

Villeneuve, J.P., Holm, E. & Cattini C. (1985). Transfer of chlorinated hydrocarbons in the food chain–reindeer–man. *Chemosphere* 14, 1651–8.

Villeneuve, J.P., Fogelqvist, E. & Cattini, C. (1988). Lichens as bioindicators for atmospheric pollution by chlorinated hydrocarbons. *Chemosphere* 17, 399–403.

von Arb, C. & Brunold, C. (1989). Lichen physiology and air pollution. I. Physiological responses of in situ *Parmelia sulcata* among air pollution zones within Biel, Switzerland. *Canadian Journal of Botany* 68, 35–42.

Walthier, D.A., Ramelow, G.J., Beck, J.N., Young, J.C., Callahan, J.D. & Marcon, M.F. (1990). Temporal changes in metal levels of the lichens *Parmotrema praesorediosum* and *Ramalina stenospora*, southwest Louisiana. *Water, Air and Soil Pollution* 53, 189–200.

Wetmore, C. M. (1989). Lichens and air quality in Cuyahoga Valley National Recreation Area, Ohio. *Bryologist* 92, 273–81.

Will-Wolf, S. (1988). Quantitative approaches to air quality studies. In *Lichens, Bryophytes and Air Quality*, ed. T.H. Nash & V. Wirth, pp.109–40. Berlin: J. Cramer.

Xue, H.B., Stumm, W. & Sigg, L. (1988). The binding of heavy metals to algal surfaces. *Water Research* 22, 917–26.

Zakshek, E.M. & Puckett, K.J. (1986). Lichen sulphur and lead levels in relation to deposition patterns in eastern Canada. *Water, Air and Soil Pollution* 30, 161–9.

Useful addresses

Information on current lichen research and on lichenologists in various countries can be obtained via the International Lichenological Newsletter, published three times a year. This is the vehicle through which lichenologists exchange news and ideas. Write to either of the editors: Dr H.J.M. Sipman, Botanische Garten & Botanische Museum, Königin-Luise Strasse 6–8, D-1 Berlin 33, Germany; or Professor M.R.D. Seaward, School of Environmental Science, University of Bradford, Yorkshire BD7 1DP, UK.

Many countries have lichen societies, and some, such as the British Lichen Society, run workshops and field meetings that welcome students and amateurs including beginners. These societies may be contacted at the following addresses:

Australasia: Society of Australasian lichenologists. Dr J.A. Elix, Department of Chemistry, the Australian National University, GPO Box 4, Canberra ACT 2601, Australia.

Central Europe: Bryologisch–Lichenologische Arbeitsgemeinschaft für Mitteleuropa (BLAM). Dr G. Philippi, Landessammlungen für Naturkunde, Erbprinzenstrasse 3, Postfach 3949, D-7500 Karlsruhe, Germany.

Czechoslovakia: Bryological and Lichenological Section of the Czechoslovak Botanical Society. Dr I. Novotny, Botanicke odd. Moravskeho Muzea, Preslova 1, CS-60200 Brno, Czechoslovakia.

Finland: Lichen section, Societas Mycologica Fennica. Dr T. Ahti, Department of Botany, University of Helsinki, Unioninkatu 44, SF-00170 Helsinki, Finland.

France: Association Française de Lichenologie (AFL). Dr R. Lallement, Université de Nantes, Laboratoire de Biologie et Cytophysiologie Végétales, 2 Rue de la Houssinière, F-44072 Nantes Cédex, France.

Italy: Societa Lichenologica Italiana (SLI). Professor G. Caniglia, Departimento di Biologia, Via Orto Botanico 15, I-35123 Padova, Italy.

Japan: Lichenological Society of Japan (LSJ). Dr H. Kashiwadani, National Science Museum, Division of Cryptogams, Hyakunin-cho 3-23-1, Shinjuku-ku, Tokyo, Japan.

Netherlands: Bryologische en Lichenologische Werkgroep der KNNV (BLW). Dr P. Hovekamp, Eiberoord 3, NL-2317 XL Leiden, The Netherlands.

Nordic Countries: Nordisk Lichenologisk Forening (NLF). Dr U. Sochting, Institut fur Sporeplanter, Ø. Farimagsgade 2D, DK-1353, København K, Denmark.

Poland: Lichenological Section of the Polish Botanical Society (Polski Towarzystwo Botaniczne). Dr W. Faltynowicz, Department of Plant Ecology, University of Gdansk, ul. Czlgistow 46, 81-378 Gdynia, Poland.

Spain: Sociedad Española de Liquenologia (SEL). Dr A. Gomez-Bolea, Departamento de Biología Vegetal (Botanica). Facultad Biología, Universidad de Barcelona, Avda Diagonal 645, 08071 Barcelona, Spain.

Switzerland: Schweizerische Vereinigung für Bryologie und Lichenologie (SVBL). Dr K. Ammann, Systematische-Geobotanisches Institut der Universität Bern, Altenbergrain 21, CH-3013 Bern, Switzerland.

UK and Ireland: British Lichen Society. Dr O.W. Purvis, Botany Department, The Natural History Museum, Cromwell Road, London SW7 5BD, UK.

USA and Canada: American Bryological and Lichenological Society (ABLS). Dr R.S. Egan, Biology Department, University of Nebraska, Omaha, NE 68182-0072, USA.

Field courses dealing with lichens and suitable for teachers and beginners are also run every year by the *Field Studies Council*. Information on these courses, which usually last a week, can be otained by writing to: Central Services (FSC), Preston Montford, Montford Bridge, Shrewsbury, Shropshire SY4 1HW.

Index

Acarospora sinopica 33
accumulation of substances 1
acid rain 13
adenosine triphosphate (ATP) 36
agrochemicals 11, 23, 30, 36
air quality zones 1, 8, 9
air pollution, fall in 1, 10, 13
algae, effects of pollution on 7, 15
ammonia 17
Anaptychia 55
  ciliaris 45
anion uptake 3
Arthonia impolita 8
aromatic hydrocarbons 22
ash weight 32, 62

bark
  cores 34
  pH 12, 14
barbed wire 28
Biston betularia 40
bisulphate 6
biological monitors; books on 65
blue-green algae (cyanobacteria) 3
bird perches; lichens of 17
Bryoria fuscescens 8, 44
Buellia punctata 9, 52
buffering capacity; 12, 14, 31

C (sodium hypochlorite) 55
cadmium 30, 62
caesium 37, 38
calcium 26
Calicium viride 8
Caloplaca 32
cannon balls 28
carbohydrate transfer 6
carboxylic acids 25
Cetraria 33
Chaenotheca ferruginea 8
chemical tests 4, 43, 52
Chernobyl 37, 38
Chiodecton 28
chloride, in sea spray 21
chlorinated hydrocarbons 22
chlorophyll
  breakdown 36
  content 6, 7, 60
  extraction 60
chromium 24, 35, 36
Chrysothrix candelaris 8
cilia 55
ciliates 42
Cladonia 27, 37, 38, 44, 56
  arbuscula 17
  ciliata 17
  coniocraea 44
  macilenta 44
  mitis 15

portentosa 17
pyxidata 44
rangiferina 14, 15, 22, 23, 25, 32, 39
squamosa 44
stellaris 14, 37
Clean Air Acts 10, 41
colorimetric techniques 62
computers, use of 10, 11, 59
contaminated water 36
cortex 55
copper 15, 24, 27, 28, 29, 33, 36, 62
cyanobacteria 3, 14

damage symptoms 60
  fluorides 18
  sulphur dioxide 10
Degelia 47
  plumbea 48
Desmococcus viridis 8, 43
digestion of lichens 62
Dimerella lutea 8
Diploicia canescens 51
distribution surveys 57
dust 1, 11, 24, 27, 32
dry deposition 15, 17, 37

electrolyte leakage 7
Euglypha species 42
eutrophication 17
Evernia prunastri 8, 9, 10, 17, 18, 46

fallout
  atmospheric 24
  from long distance transport 33
Field Studies Council 5, 73
fluorides 18
foliose 55
Foraminella ambigua 8, 49
fruit bodies 55
  round 4, 53
  elongate 54
fruticose 55

geobotanical prospecting 32
geothermal exploration 31
Glaucoma scintillans 42
Graphis 51
  elegans 8, 51
  scripta 51
gravestones 28
growth rates 3

half-life, biological 38
Hawksworth and Rose scale 1, 8, 9
heathlands 14
herbicides 23
Heterodermia 28
highways 30, 31
Humerbates rostrolamellatus 41
hydrogen fluoride 21
Hypocenomyce scalaris 8

Hypogymnia 47
  physodes 8, 9, 13, 15, 18,
    19, 20, 30, 33, 34, 35, 48, 60

Indices of Atmospheric Purity,
  IAPs 10, 19, 59
  identification 4, 5, 43, 57
  insecticides 23
  invertebrates 4, 40
  ion exchange 3, 24, 31
  iron 27, 29, 33, 38, 62
  isidia 4, 55

K (potassium hydroxide) 55

Lapps 37, 38
lead 24, 27, 28, 30, 31, 61, 62
Lecanora 51
  cascadensis 28, 33
  chlarotera 8, 51
  conizaeoides 8, 9, 17, 29, 42, 43, 51
  dispersa 9, 11, 51
  expallens 8, 19, 44, 51
  muralis 12
  piniperda 9
  polytropa 33
  vinetorum 28
Lepraria
  candelaris 44
  incana 8, 9, 17, 44, 52
  lobificans 44
leprose 55
lichen
  acids 4, 19
  desert 1
  distribution 3, 8, 57
  nitrogen content 17
  societies 73
  substances 4, 19
  zones 1, 8, 9
Lecidea lapicida 32
Lecidella elaeochroma 52
Lobaria
  amplissima 8, 14
  pulmonaria 8, 47
  scrobiculata 8, 14, 47
  lobules 55

Macrobiotus hufelandii 42
magnesium 26
mapping 13
medulla 55
melanism 40, 57, 63
menstruation 38
mercury 29, 31, 36, 62
metals 24, 61
  airborne 35
  binding affinity 25
  classification 26
  toxicity 26
microfauna 41, 63, 66

Milnesium tardigradum 42
mines 31, 60
mites 46
moths 62
Mycorrhizae 39

National Parks 13
nickel 15, 29, 36, 61
nitrate 14, 17
nitric acid 14
nitrogen fixation 3
nitrogen oxides 14, 17
nitrogen compounds 17
Normandina pulchella 8
Nostoc 54

oil refineries 27
Opegrapha
 atra 51
 varia 51
oxalates 28
ozone 16

P (paraphenylenediamine) 55
Pachyphiale cornea 8
Pannaria 8, 47
 rubiginosa 48
Parmelia 28, 30, 47, 50, 55
 acetabulum 49
 baltimorensis 30, 31
 caperata 8, 9, 18, 19, 34, 49
 exasperatula 8
 glabratula 8, 9, 49
 loxodes 20
 pastillifera 49
 perlata 8, 19, 48
 praesorediosa 29, 36
 reticulata 8
 revoluta 8
 saxatilis 8, 9, 18, 19, 49
 subaurifera 49
 sulcata 8, 9, 18, 19, 30, 31, 48
 subrudecta 8, 19, 49
 tiliacea 8, 49
particulates 1, 24, 27, 32
PCBs (polychlorinated
 biphenyls) 22
Peltigera 27, 28, 33, 56
Penicillium notatum 26
Pertusaria
 albescens 8
 amara 8
 hemisphaerica 8
 hymenea 8
pesticides 23
pH 12, 14
Phaeophyscia 47, 50
 orbicularis 11, 50
Physcia 17, 32, 47, 50, 55
 adscendens 50
 aipolia 50

caesia 50
leptalea 45
tenella 50
Physconia 47, 50
 distorta 50
 enteroxantha 50
Platismatia glauca 8, 9, 47
plutonium 37, 38
podetium 55
Porpidia macrocarpa 33
potassium 7, 14
Pseudevernia furfuracea 8, 20, 46
Pseudocyphellaria 47
pulp mills 3

quarries 32, 60

radioactive elements 37
radionuclides 2
Ramalina 9, 28
 calicaris 46
 canariensis 46
 duriaei 26, 35, 36
 farinacea 8, 46
 fastigiata 9, 46
 fraxinea 46
 menziesii 16
 siliquosa 18, 19
regression model 63
reindeer 4, 23, 37, 38
report preparation and
 publication 64
reproduction 4
resynthesis 4
rhizinae 4, 28, 55
rhizopods 42
Rinodina
 gennarii 11
 roboris 8
rocks, damage to lichens on 19

satellites 38
school pupils 9, 57
Scoliciosporum
 chlorococcum 9
 umbrinum 11
sea spray 15, 21
smelters
 aluminium 18
 copper 15
 iron 29, 35
 nickel 12, 15, 27
soredia 4, 55
spores 43, 53, 55
Sticta 47
 limbata 8
sulphate 14
sulphite 6
sulphur 15, 34
 accumulation 15
sulphur dioxide 6

symptoms of damage 10
 toxic effects 6
sulphuric acid 14, 34
sulphurous acid 6
Stereocaulon 29
 vesuvianum 21

tardigrades 42
telephone wires 28
television aerials 28
Teloschistes flavicans 8, 28, 44
titanium 27
thallus 55
transplanting lichens 34
Trebouxia 14, 54
trees
 damage from lichens 18
 marking 59
 selection 58
Tremolecia atrata 32
Trinema lineare 42

Umbilicaria 39
 muhlenbergii 25, 26
uptake mechanisms
 anions 3
 metals 26
 particulates 27
urban areas 34, 57
uranium 32, 39
Usnea 11, 23, 57
 articulata 8, 45
 ceratina 8
 florida 8, 45
 filipendula 8
 rubicunda 8, 45
 subfloridana 8, 11, 45

vanadium 27
volcanoes 21

water loss 3
Woodruffia lichenicola 42

Xanthoria 11, 17, 32
 candelaria 46
 parietina 21, 46
 polycarpa 11, 46

zinc 25, 36, 62
zones, air quality 1, 8, 9

Printed in the USA
CPSIA information can be obtained
at www.ICGtesting.com
JSHW010604291024
72561JS00010B/40

9 781784 272111